BIOTERRORISM

BIOTERRORISM

Ernest P. Chiodo,
M.D., J.D., M.P.H., M.S., M.B.A., C.I.H.

Physician-Attorney
Diplomate of the American Board of Internal Medicine
Diplomate of the American Board of Preventive Medicine
in Occupational Medicine
Diplomate of the American Board of Preventive Medicine
in Public Health and General Preventive Medicine
Diplomate of the American Board of Industrial Hygiene as a
Certified Industrial Hygienist in the Comprehensive
Practice of Industrial Hygiene

Library of Congress Control Number:		2013901403
ISBN:	Hardcover	978-1-4797-8431-8
	Softcover	978-1-4797-8430-1
	Ebook	978-1-4797-8432-5

This book was printed in the United States of America.

To order additional copies of this book, contact:
Xlibris Corporation
1-888-795-4274
www.Xlibris.com
Orders@Xlibris.com
129829

CONTENTS

ABOUT THE AUTHOR

E RNEST P. CHIODO is a physician as well as an attorney licensed to practice medicine and law in Michigan and Illinois. He is a graduate of Kalamazoo College, Wayne State University School of Medicine, Wayne State University Law School, and Harvard University School of Public Health. He also has a Master of Science in Threat Response Management (biological, chemical, and radiological defense) from the University of Chicago, a Master of Science in Biomedical Engineering from Wayne State University, and a Master of Science in Industrial Toxicology from Wayne State University. He also has a Master of Business Administration with a concentration in economics from the University of Chicago. He is board certified in the specialties of internal medicine, occupational and environmental medicine, public health and general preventive medicine, and industrial hygiene. He has served as the Medical Director of the Detroit Health Department and as the Medical Director of the pension boards of the City of Lansing, Michigan. He is a clinical assistant professor of family medicine and public health at Wayne State University School of Medicine. He has also served as an adjunct professor of industrial hygiene and industrial toxicology at Wayne State University. His is an adjunct professor of law at John Marshall Law School and at Loyola University Law School in Chicago where he teaches toxic tort law. He has clinical privileges in the Henry Ford Health System and maintains a medical practice in occupational and environmental medicine. In addition to his medical practice, he has an active legal practice in Michigan and Illinois. He has served as the Chairman of the Environmental Litigation and Administrative Practice Committee of the Environmental Law Section of the State Bar of Michigan. He has been elected by his fellow environmental attorneys to serve on the Environmental Law Council of the State Bar of Michigan. He is a past president of the Michigan Industrial Hygiene Society. He has also served as the international advisor to Goodrich Corporation in their preparations for pandemic influenza.

INTRODUCTION

T HIS BOOK IS about agents that may be used in acts of biological, chemical and radiological terrorism. The material is presented at a level understandable by an educated person with a basic understanding of medical concepts. The material is divided into five sections. The first section provides an overview of the types of technical experts that may be of help in preparing and responding to acts of biological, radiological, and chemical terrorism. The second section is focused upon biological terrorism agents. The third section is focused upon chemical terrorism agents. The fourth section is concerned with radiological agents that may be utilized in terrorism. The fifth section provides an overview of technical issues that are of importance in the understanding of biological, chemical, and radiological terrorism. The overview of technical issues includes brief overviews of toxicology and epidemiology. The final section of the first part of this book covers legal issues concerning bioterrorism.

Terrorism is the intentional creation of fear in populations in order to achieve an objective. That objective may be religions, political, or financial. It is the goal of the terrorist to create anxiety that is far greater than the actually physical damage caused by the terrorist act. While terrorism is frequently enacted in order to effect political change, the targets of terrorism include governmental facilities, government officials, citizens, as well as corporations.

TECHNICAL EXPERTS AND TRAINING PROGRAMS

TERRORISM TECHNICAL EXPERTS

*More is experienced in one day in the life of a learned
man than in the whole lifetime of an ignorant man.*

Seneca

THE DIFFERENCE BETWEEN success and failure in response to terrorism is technical expertise. There are many types of experts that come to play a role in issues of biological, chemical and radiological terrorism. The technical experts include physicians with various areas of specialization; industrial hygienists; epidemiologists; toxicologists; and engineers. The list goes on and on. This book can not discuss all the various types of experts that may be called upon to prepare for or respond to acts of terrorism. However, the strengths and weaknesses of the most frequently utilized experts will be considered.

Physician

THE ADVICE OF qualified physicians is vital to prepare for and respond to threats of biological, chemical and radiological terrorism. However, most non-physicians have little or no knowledge of the range of expertise covered by various areas of medical specialization. There are over 800,000 physicians in the United States. Many of these physicians are board certified in some specialty; however, what does it mean to be board certified? When the term "board certified" is used in reference to a physician, it is usually meant that the physician is certified by one of the American Board of Medical Specialties (ABMS) boards. These are the boards that hospitals consider when determining whether a physician has proper credentials to serve on staff. The ABMS boards consist of twenty-three specialty boards including: the American Board of Internal Medicine; the American Board of Surgery; the American Board of Pediatrics; and the American Board of Preventive Medicine. Physicians who are board certified by the American Board of Preventive Medicine are likely to be central in any preparation or response to terrorism. The American Board of Preventive Medicine grants board certification in three areas of specialization. These are (1) aerospace medicine; (2) public health and general preventive medicine; and (3) occupational medicine. Specialists in aerospace medicine are the types of physicians that serve on the staff of NASA. Specialist in public health and general preventive medicine are the types of physicians that serve as medical directors of state and local health departments or work for the Centers for Disease Control (CDC). Specialists in occupational medicine (occupational and environmental medicine) are the physicians that specialize in the diagnosis, treatment, and prevention of diseases caused by work place and environmental exposures. This includes exposure to toxic substances at work or in the environment.

Board certification in public health and general preventive medicine or in occupational and environmental medicine requires formal education beyond medical school in the form of a Master of Public Health. In public health school, the board certified occupational and environmental medicine physician has usually received formal training in biostatistics, epidemiology, and toxicology. Knowledge of these disciplines is essential for the preparation and response to biological, chemical and radiological terrorism. This knowledge is not emphasized during the training of most physicians.

Epidemiology is the discipline involved in surveillance of disease in populations that may be the first sign of a terrorist attack. Public health and general preventive medicine is the epidemiological discipline within medicine.

A well-qualified occupational and environmental medicine physician will have formal training and academic course work in toxicology. Consequently, issues concerning toxicology are often within the expertise of a board certified occupational and environmental medicine physician.

Physicians who are board certified in emergency medicine are also central to preparation and response to acts of biological, chemical, and radiological terrorism. Emergency medicine physicians are usually the first physicians that are likely to see victims of terrorism. They possess special training and expertise in the emergency treatment of victims of terrorism. They also have specialized knowledge of implementation of systems including deployment of first responders to facilitate the most effective treatment of victims of biological, chemical, and radiological terrorism while protecting the safety of first responders and other emergency and medical personnel.

Industrial Hygienists

INDUSTRIAL HYGIENISTS ARE the professionals that conduct sampling and monitoring of toxic substances in the workplace and the environment. The American Board of Industrial Hygiene grants certification to industrial hygienists with the designation of C.I.H. (Certified Industrial Hygienist). Most industrial hygienists come into the field with a background in an engineering or chemical discipline. The knowledge of an industrial hygienist may be vital in issues of biological, chemical and radiological terrorism. The Certified Industrial Hygienist is the central expert in issues of monitoring and controlling exposures to chemicals, biological, and radiological agents that may be used by terrorists.

Other Experts

EXPERTS IN OTHER disciplines are commonly needed in matters concerning biological, chemical and radiological terrorism. Psychologist and psychiatrists maybe necessary to address the mental health needs of victims of terrorism. Various engineering and architectural experts may be necessary to address the risk to buildings and infrastructure from terrorism. A hydrologist is frequently needed in order to describe and quantify the risk of ground water contamination. A civil and environmental engineer may be required to prevent attacks on municipal water supplies.

TRAINING PROGRAMS

Training is everything. The peach was once a bitter almond;
cauliflower is nothing but cabbage with a college education.

Mark Twain

TRAINING IN THE technical and administrative aspects of biological, chemical, and radiological terrorism is necessary for the practitioner to formulate effective counter measures. The necessary training can be obtained in a number of different venues. Of course, the military is one source of counter terrorism training. Expertise can also be gained through medical schools, schools of public health, schools of public administration, and well as in other undergraduate and graduate programs dealing with threat response management or its sub-disciplines. The University of Chicago has recently started a program leading to a Master of Science in Threat Response Management. This program is housed within the Graham School of General Studies at the Gleacher Center located at 450 N. Cityfront Plaza Drive in downtown Chicago just east of Michigan Avenue along the Chicago River.

In order to earn the Master of Science degree in Threat and Response Management the student must take nine courses in addition to completing a capstone or practicum project. The program is divided into a scientific and administrative tract. Students in the first year take the same courses. The courses during the first trimester includes the *Foundations of Emergency Management and Homeland Security* and *Analyzing and Communicating Public Policy, Legal, and Ethical Issues.* The courses during the second trimester are *Biostatistics* and *Leadership and Management.* The courses during the third trimester of the first year are *Public Health Investigation and Surveillance—tracking the Health of Populations* and *Psychological, Social, and Behavioral Contexts of Emergency and Hazard Response.* During the second year students in the scientific track take *Epidemiology and Infectious Disease, Ionizing Radiation and Chemical Hazards—Pathobiology, Toxicology, Biohazards* and *The Vectors of Disease—Food, Water, Air, Insects.* During the second year students in the administrative track take *Complex Adaptive Systems for Emergency Preparedness and Homeland Security, Technology strategy and Information Systems, and Financial and Resource Planning for Risk and Crisis Management.*

BIOLOGICAL AGENTS OF TERROR

*Biology occupies a position among the sciences both marginal and
central. Marginal because, the living world constituting only a tiny
and very 'special' part of the universe, it does not seem likely that the study
of living beings will ever uncover general laws applicable
outside the biosphere. But if the ultimate aim of the whole of science
is indeed, as I believe, to clarify man's relationship to the universe,
then biology must be accorded a central position, since of all the
disciplines it is the one that endeavors to go most directly to the heart
of the problems that must be resolved before that of 'human nature'
can even be framed in other than metaphysical terms.*

Jacques Lucien Monod

B IOLOGICAL AGENTS OF terror include bacteria, protozoa, fungi, viruses, and prions. Biological agents also include bacterial and fungal toxins. The incomplete participation by several signatory nations to the 1972 Biological Weapons Convention treaty resulted in development of scientific expertise in bioterrorism agents with the support of nation states.[1] The ideal bioterrorism agent is easy and inexpensive to produce from readily available materials, lethal or seriously incapacitating, stable in an aerosolized form, easy to disperse over wide areas, resistant to degradation in the environment, have no effective vaccine or anti-dote, and be communicable from person to person.[2] Bioterrorism agents have the potential of causing deaths in numbers comparable to those caused by nuclear weapons. Biological agents used as terrorist weapons have the potential of causing a general collapse of the fabric of society due to great loss of life including human and livestock, famine, and economic disruption.[7, 8, 9]

The appeal of biological agents as terrorist weapons include their ease of production and dispersal, their delayed onset, high mortality and morbidity rates, and difficulty of detection, diagnosis, and treatment.[7, 8] For example, a terrorist can obtain viable spores of anthrax in the soil from many agricultural areas throughout the world. Culturing and growth of anthrax is relatively simple and requires equipment that is easily obtained or manufactured by the general public. Disbursal of anthrax spores can occur easily and can utilize the infrastructure of modern society such as the postal system. The manifestation of disease due to anthrax is delayed for at least a few days after exposure. This makes the identification and apprehension of terrorists difficult.

The history of military use of bio-warfare agents going back to catapulting corpses of plague victims into the city of Kaffa during the middle ages combined with the ease of dispersal of biological warfare agents as contaminants of various environmental media such as air, water, food, soil, and fomites has lead many governments to prioritize plans to respond to bioterrorist attacks.[7, 8, 10, 11, 12]

Toxins are toxic chemicals that are formed in nature by living organisms including bacteria, fungi, plants, and animals. Toxins include the most potent poisons on the Planet. However, factors such as ease of production, stability in the environment, stability during and after delivery attempts such as aerosolization, and toxin potency after weaponization attempts are factors that influence the use of toxins as agents of bioterrorism. The two most important toxins from the perspective of potential bioterrorism agents are botulinum toxin and staphylococcal enterotoxin B (SEB). Other toxins that are potential bioterrorism agents are ricin and mycotoxins including aflatoxins and trichothecenes.[3]

The Federal Working Group for Civilian Biodefense has created a list of biological agents that make them particularly effective as weapons.[4, 5] These features are as follows:

1. *High morbidity and mortality;*
2. *Potential for person-to-person spread;*
3. *Low infective dose and highly infectious by aerosol;*
4. *lack of rapid diagnostic capability;*
5. *Lack of universally available effective vaccine;*
6. *Potential to cause anxiety;*
7. *Availability of pathogen and feasibility of production;*
8. *Environmental stability;*
9. *Database of prior research and development;*
10. *Potential to be "weaponized".*

The U.S. Centers for Disease Control and Prevention (CDC) has classified three categories of potential biological threats based upon the priority of the pathogen's threat. The highest priority pathogen are Category A agents. These agents present the greatest risk to national security since they result in the highest mortality rates and present a major potential public health impact; they are easily transmitted from person to person; they are capable of causing public panic and social disruption; they require special preparations to be made for public health protection. Category B agents are ranked next in priority due to their moderate ease in dissemination; their low mortality and moderate morbidity rates; they require special diagnostic procedures. Category C agents are ranked third in priority. These agents include emerging pathogens to which the general public lacks immunity. These agents may also be suitable to modification through biomedical engineering to increase their risk potential.[4]

The Category A Agents are as follows:[4, 5]

Bacillus anthracis (Anthrax)
Clostridium botulinum toxin (Botulism)
Yersinia pestis (Plague)
Variola major (Smallpox)
Francisella tularensis (Tularemia)
Viral hemorrhagic fevers (Arenaviruses, Bunyaviridae, Filoviridae,
                    And Flaviviridae)

The Arenaviruses include Lassa fever virus and the viruses causing South American hemorrhagic fever including Junin, Machupo, Guanarito, and Sabia viruses.

The Bunyaviridae viruses include those viruses causing Crimean-Congo and Rift Valley hemorrhagic fevers.

Filoviridae include the Ebola and Marburg viruses.

Flaviviridae include the Yellow fever, Omsk fever, and Kyasanur Forest viruses.

The Category B Agents are as follows:[4, 5]

Brucella species (Brucellosis)
Epsilon toxin of *Clostrdium perfringens*
Yersinia pestis (Plague)
Salmonella species, *Escherichia coli* 0157:H7, *Shigella* and other threats
to food safety
Burkholderia mallei (Glanders)
B. pseudomallei (Meliodosis)
Chlamydia psittaci (Psittacosis)
Coxiella burnetii (Q fever)
Ricinus communis toxin (Ricin from castor beans)
Staphylococcal enterotoxin B
Rickettsia prowazekii (Typus fever)
Viral encephalitis viruses
Vibrio cholerae, Cryptosporidium parvum and other water safety threats

The Category C Agents are as follows:[4, 5]

Hantavirus, Nipah, SARS coronovirus and other emerging infections diseases

1. Zapor M, Fishbain JT. Aerosolized biologic toxins as agents of warfare and terrorism. Respir Care Clin N Am. 2004 Mar;10(1):111-22.
2. Henghold WB 2nd. Other biologic toxin bioweapons: ricin, staphylococcal enterotoxin B, and trichothecene mycotoxins. Dermatol Clin. 2004 Jul;22(3):257-62, v.
3. Madsen JM. Toxins as weapons of mass destruction. A comparison and contrast with biological-warfare and chemical-warfare agents. Clin Lab Med. 2001 Sep;21(3):593-605.
4. Harrison's Principles of Internal Medicine. 16th Edition. © 2005 by the McGraw-Hill Companies. New York. Page 1279.
5. Borio L, et al. JAMA 287:287:2391, 2002.
6. Hart T. Microterrors: the complete guide to bacteria, viral and fungal infections that threaten our health. © 2004 by Axis Publishing Ltd. Buffalo, New York. Pages 47 and 48.
7. Sinclair R, Boone SA, Greenberg D, Keim P, Gerba CP. Persistence of category A select agents in the environment. Appl Environ Microbiol. 2008 Feb;74(3):555-63.
8. Broussard LA. 2001. Biological agents: weapons of warfare and bioterrorism. Mol. Diagn. 6:323-333.
9. Stuart AL. Wilkening DA. 2005. Degradation of biological weapons agents in the environment: implications for terrorism response. Environ. Sci. Technol. 39:2736-2743.
10. Henderson DA. 2002. Bioweapons preparedness chief discusses priorties in world of 21st-century biology. Interview by Rebecca Voelker. JAMA 287:573-575.
11. Henderson DA, Inglesby TV, Bartlett JG, Ascher MS, Eitzen E, Jahrling PB, Hauer J, Layton M, McDade J, Osterholm MT, O'Toole T, Parker G, Perl T, Russel PK, Tonat K, et al. 1999. Smallpox as a biological weapon: medical and public health management. JAMA 281:2127-2137.
12. Holdstock D. 2000. Biotechnology and biological warfare. Peace Rev. 12:549-553.

TOXINS

*The specific character of the greater part of the toxins which
are known to us (I need only instance such toxins as those of
tetanus and diphtheria) would suggest that the substances
produced for effecting the correlation of organs within the
body, through the intermediation of the blood stream, might
also belong to this class, since here also specificity of action
must be a distinguishing characteristic.*

Ernest Henry Starling

BOTULINUM TOXIN

Shape without form, shade without color, paralyzed force,
gesture without motion; those who have crossed with
direct eyes, to death's other kingdom remember us.

Thomas Stearns Eliot

B OTULISM IS A serious paralyzing disease caused by a toxin produced by the bacteria Clostridium botulinum. The toxin made by Clostridium botulinum causes paralysis of the skeletal muscle through a presynaptic blockage of the release of the neurotransmitter acetylcholine.[1] Botulinum toxin causes death through paralysis of the respiratory muscles.[8,9] Botulinum toxin is the most potent poison known.[1,2,3] The potency of botulinum toxin as a poison is highlighted by the fact that one gram of crystalline botulinum toxin if inhaled could kill one million persons.[1] Unit 731, the Japanese biological warfare unit, experimented with botulinum toxin on prisoners in Manchuria during World War II.[1,4] Aum Shinrikyo, a Japanese cult known for its terrorist activities, attempted to disperse aerosolized botulinum toxin in downtown Tokyo on at least 3 occasions between 1990 and 1995. These attacks failed due to defective aerosolization equipment and improper microbiological technique.[1,5,6]

There are seven types of botulinum toxin that are designated with letters from A through G.[1] Of the seven types of botulinum toxin, type A has the highest mortality rate and consequently has the greatest potential for use as a bioterrorism agent.[7]

Botulinum toxin causes a descending flaccid paralysis and dysfunction of the nervous system.[10] Victims of botulinum intoxication suffer from dysfunction of speaking, seeing, and swallowing. They typically have double vision (diplopia), blurred vision, drooping of the eye lids (ptosis), and dilated or sluggish pupils.[1,11,12,13,14] Victims also suffer from mouth dryness due to blockade of the parasympathetic nervous system. They also suffer from loss of control of their head due to muscular weakening of the supporting musculature. There is loss of the gag reflex so that oral secretions can be aspirated leading to possible serious mixed anaerobic pneumonia.[1]

Botulinum toxin is odorless, colorless, and has no taste.[1] This is a factor that would also assist a stealth bioterrorist attack through contamination of food. However, botulism toxin is readily inactivated by heat equal or greater to 85°C for 5 minutes.[1,15,16,17]

1. Cherington M. Clinical spectrum of botulism. Muscle Nerve. 1998 Jun;21(6):701-10.
2. Gill MD. Bacterial toxins: a table of lethal amounts. Microbiol Rev. 1982; 86-94.
3. National Institute of Occupational Safety and Health. Registry of Toxic Effects of Chemical Substances (R-TECS). Cincinnati, Ohio: National Institute of Occupational Safety and Health; 1996.
4. Hill EV. Botulism. In: Summary Report on B.W. Investigations. Memorandum to Alden C. Waitt, Chief Chemical Corps, United States Army, December 12, 1947; tab D. archived at the US Library of Congress.
5. Tucker JB, ed. Toxic Terror: Assessing the Terrorist Use of Chemical and Biological Weapons. Cambridge, Mass: MIT Press; 2000.
6. WuDunn S, Miller J, Broad WJ. How Japan germ terror alerted world. New York Times. May 26, 1998: A1, A10.
7. Shukla HD, Sharma SK. Clostridium botulinum: a bug with beauty and weapon. Crit Rev Microbiol. 2005;31(1):11-8.
8. Josko D. Botulin toxin: a weapon in terrorism. Clin Lab Sci. 2004 Winter;17(1):30-4.
9. Patocka J, Splino M, Merka V. Botulism and bioterrorism: how serious is this problem? Acta Medica (Hradec kralove). 2005;48(1):23-8.
10. Caya JG. Clostridium botulinum and the ophthalmologist: a review of botulism, including biological warfare ramifications of botulinum toxin. Surv Opthalmol. 2001 Jul-Aug;46(1):25-34.
11. Franz Dr, Jahrling PB, Friedlander Am, et al. Clinical recognition and management of patients exposed to biological warfare agents. JAMA. 1997;278:399-411.
12. Hughes JM, Blumenthal JR, Merson MH, Lombard GL, Dowell VR Jr, Gangarosa EJ. Clinical features of types A and B food-borne botulism. Ann Intern Med. 1981;95:442-445.
13. Shapiro RL, Hatheway C, Swerdlow DL. Botulism in the United States: a clinical and epidemiologic review. Ann Intern Med. 1998;129:221-228.
14. Middlebrook JL, Franz DR. Botulinum toxins. In: Sidell FR, Takafuji ET, Franz DR, eds. Medical Aspects of Chemical and Biological Warfare. Washington, DC: Office of the Surgeon general; 1997:643-654. Textbook of Military medicine; part 1, vol 3.
15. Smith LDS. Botulism: The Organism, Its Toxins, the Disease. Springfield, IL: Charles C. Thomas Publisher; 1977.

16. Hatheway CL, Johnson EA. Clostridium: the spore-bearing anaerobes. In: Collier L, Balows A, Sussman M, eds. Topley & Wilson's Microbiology and Microbial Infections. 9th ed. New York, NY: Oxford University Press; 731-782.
17. Siegel LS. Destruction of botulinum toxin in food and water. In: Hauschild AH, Doods KL, eds. Clostridium botulinum: Ecology and Control in Foods. New York, NY: Marcel Dekker Inc; 1993:323-341.

RICIN

I will arise and go now, and go to Innisfree,
And a small cabin build there, of clay and wattles made:
Nine bean-rows will I have there, a hive for the honeybee,
And live alone in the bee-loud glade.

William Butler Yeats

RICIN IS A protein plant toxin (toxalbumin) 60 to 65 kilo-Daltons derived from beans of the castor bean plant, *Ricinus communis*.[1, 5] *Ricinus communis* is a member of the Euphorbiaceae family of plants.[16] Caster beans have an oblong shape and are light brown in color with dark brown spots.[2] In additions to serving as a laxative, caster bean oil is also used as a lubricant in high speed automobiles, industrial machinery, and jet engines as well as a component in paints and varnishes.[2, 10] Over one million tons of castor beans are harvested annually.[20] Ricin is a purified white powder that is stable over a wide pH range and is soluble in water.[2, 3, 4] Ricin is inactivated by heating for one hour in an aqueous solution at 80°C.[2, 4] However, longer periods of heating are required when ricin is in powdered form.[2] Aqueous solutions of ricin are resistant to chlorine at 10 ppm.[17] Caster beans have a content of ricin ranging between 1 and 5 percent.[2, 5, 6] Ricin is contained in the pulp of the caster beans after removal of the caster oil.[2] Ricin is easier to produce than either botulinum toxin or anthrax.[20]

Ricin has been considered as a possible chemotherapeutic agent since cancer cells have more carbohydrate containing surface lectin binding sites mediating ricin toxicity.[2, 7, 8, 9] Ricin acts enzymatically to block protein synthesis by inducing hydrolytic fragmentation of ribosomes.[15] Ricin binds to carbohydrates on cell surfaces and then is internalized in the cell leading to cell death by inhibiting protein synthesis.[21] Ricin enters cells by binding through the ricin B chain to cell surface glycolipids and glycoproteins that have beta 1,4 linked galactose residues. Following this glycolipid or glycoprotein binding, ricin is taken into the cells through endocytosis. Ricin is then taken into endosomes. Ricin then is taken from endosomes into the trans-Golgi network. Ricin is taken from the Golgi complex into the endoplasmic reticulum. In the endoplasmic reticulum ricin interferes with ribosomal production of protein.[22]

Ricin was considered as a warfare agent by the United States War Department as early as 1918 at the end of World War I.[2, 3] Weapons grade ricin was made and fitted into artillery shells in the 1940s by the United States and the 1980s by Iraq.[2, 4, 11] In 1978 Bulgarian journalist Georgi Markov appears to have been assassinated in

London by a KBG agent.[2, 12, 13] Ricin was found in a letter addressed to the White House as well as in a mail sorting facility in South Carolina in 2003.[2, 14]

Ricin is hazardous through inhalation, ingestion, and penetration through broken skin.[17] Release of ricin as an aerosol into the environment or adulteration of food are the exposure routes mostly likely to be used by terrorists.[2] The general symptoms of ricin intoxication include fever, fatigue, weakness, as well as muscle and joint pain.[17] Symptoms that occur after ingestion with an onset within 12 hours. These symptoms are nonspecific and may include nausea, vomiting, abdominal pain, and diarrhea which may be bloody. Victims of ricin ingestion may progress to suffer from hypotension, liver failure, and renal failure. Symptoms may persist for several days until the victim either recovers or dies due to multi-organ failure or cardiovascular collapse.[2, 17] Symptoms of inhalational exposure usually have an onset within 8 hours. The symptoms of inhalation exposure to ricin include cough, shortness of breath, chest tightness, mucous membrane irritation, joint pain, and fever. Victims of inhalational ricin exposure may progress to pulmonary edema and hypoxemia (low blood oxygen) and death due to respiratory failure although multi-organ failure may occur.[2, 17] There is little published data concerning human toxicity due to injection of ricin. The LD_{50} in mice after injection is 5 to 10 μg/kg.[2, 7, 18] The clinical presentation of victims of ricin injection include nonspecific signs and symptoms that may be similar to sepsis that include headache, fever, dizziness, anorexia, nausea, hypotension, and abdominal pain. These signs and symptoms may not occur for 10 to 12 hours.[2, 7, 18, 19] It is estimated that the human lethal dose by inhalation and injection is 5 to 10 μg/kg. For a 70 kg adult this is 350 to 700 μg.[21]

RiVax, a recombinant ricin A chain vaccine is a potential protective measure against either gastrointestinal or inhalational exposure to ricin.[1] There is no available antidote for ricin.[23]

1. Smallshaw JE, Richardson JA, Vitetta ES. RiVax, a recombinant ricin subunit vaccine, protects mice against ricin delivered by gavage or aerosol. Vaccine. 2007 Oct 16;25(42):7459-69.

2. Audi J, Belson M, Patel M, schier J, Osterloh J. Ricin poisoning: a comprehensive review. JAMA. 2005 Nov 9;294(18):2342-51.

3. Cope AC, Dee J, Cannan RK, et al. Chemical Warfare Agents and Related Chemical Problems—Part I: Summary Technical Report of Division 9. Washington, DC: National Defense Research Committee. 1945; 179-203.

4. Parker DT, Parker AC, Ramachandran CK. Joint Technical Data Source Book. Vol 6. Part 3. US Dug-way Proving Ground, Utah: Joint Contact Point Directorate. 196;1-38. DGP No. DPGJCP-961007.

5. Balint GA. Ricin: the toxic protein of caster oil seeds. Toxicology. 1974;2:77-102.

6. Bradberry SM, Dickers KJ, Rice P, Griffiths GD. Vale JA. Ricin poisoning. Toxicol Rev. 2003;22;65-70.

7. Godal A, Fodstad O, Ingebrigsten K, Pihl A Pharmacological studies of ricin in mice and humans. Cancer Chemother Pharmacol. 1984;13:157-163.

8. Lord JM, Gould J, Griffiths D, et al. Ricin: cyto-toxicity, biosynthesis and use in immunoconjugates. Prog Med Chem. 1987;24:1-28.

9. Lord JM, Roberts LM, Robertus JD. Ricin: structure, mode of action, and some current applications. FASEB J. 1994;8:201-208.

10. Brugsch HG. Toxic hazards: the castor bean. N Engl J Med. 1960;262:1039-1040.

11. Zilinskas RA. Iraq's biological weapons: the past as future? JAMA. 1997;278:418-424.

12. Crompton R, Gall D. Georgi Markov: death in a pellet. Med Leg J. 1980;48:51-62.

13. Knight B. Ricin-a potent homicidal poison. BMJ. 1979;1:350-351.

14. Centers for disease Prevention and Control. Investigation of a ricin-containing envelope at a postal facility—South Carolina. 2003. MMWR Morb Mortal Wkly Rep. 2003;52: 1129-1131.

15. Casarett and Doull's Toxicology: The Basic Science of Poisons. 6th Edition. 2001. McGraw-Hill. New York. Page 46.

16. Casarett and Doull's Toxicology: The Basic Science of Poisons. 6th Edition. 2001. McGraw-Hill. New York. Page 967.

17. Ellison DH. Handbook of Chemical and biological Warfare Agents. Second Edition. © 2008 by Taylor & Fracis group, LLC. Boca Raton. Page 482.

18. Fodstad O, Olsnes S, Pihl A. Toxicity, distribution and elimination of the cancerostatic lectins abrin and ricin after parenteral injection into mice. Br J Cancer. 1976;34:418-425.

19. Fodstad O, Johannessen JV, Schjerven L, Pihl A. Toxicity of abrin and ricin in mice and dogs. J Toxicol Environ Health. 1979;5:1073-1084.

20. Doan LG. Ricin: mechanism of toxicity, clinical manifestations, and vaccine development. A review. J Toxicol Clin Toxicol. 204;42(2):201-8.

21. Bradberry SM, Dickers KJ, Rice P, Griffiths GD, Vale JA. Ricin poisoning. Toxicol Rev. 2003;22(1):65-70.

22. Lord MJ, Jolliffe NA, Marsden CJ, Pateman CS, Smith DC, Spooner RA, Watson PD, Roberts LM. Ricin. Mechanisms of cytotoxicity. Toxicol Rev. 2003;22(1):53-64.

23. Carra JH, McHugh CA, Mulligan S, Machiesky LM, Soares AS, Millard CB. Fragment-based identification of determinants of conformational and spectroscopic change at the ricin active site. BMC Struct Biol. 2007 Nov 6;7:72.

STAPHYLOCOCCAL ENTEROTOXIN B

While working with staphylococcus variants a number of culture-plates were set aside in the laboratory bench and examined from time to time. In the examinations these plates were necessarily exposed to air and they became contaminated with various micro-organisms. It was noticed that around a large colony of a contaminating mould the staphylococcus colonies became transparent and were obviously undergoing lysis. Subcultures of this mould were made and experiments conducted with a view to ascertaining something of the properties of the bacteriolytic substance which had evidently been formed in the mould culture and which had diffused into the surrounding medium. It was found that broth in which the mould had been grown at room temperature for one or two weeks had acquired marked inhibitory, bacteriocidal and bacteriolytic properties to many of the more common pathogenic bacteria.

Alexander Fleming

STAPHYLOCOCCAL ENTEROTOXIN B (SEB) is a rapidly acting cytotoxin that is a primary cause of food poisoning and is a super-antigen that can cause toxic shock.[1, 2] SEB is a white solid that is not destroyed by cooking or freezing nor is SEB destroyed by chlorine in municipal water systems. SEB can be stored for more than a year as a freeze dried powder.[1]

SEB is hazardous when either ingested or inhaled.[1] A robust systemic immune activation can occur due to conjunctival exposure to SEB from either hand to eye contact or exposure of the eye to a bioterrorist generated aerosol.[3]

SEB is a super-antigenic exotoxin produced by staphylococcus aureus that can even at extremely low concentrations produce a strong polyclonal activation of CD4+ and CD8+ T lymphocytes.[3, 4] It is the ability of SEB to cause a power immuno-stimulatory response that is the attraction of this agent as a biological weapon.[3, 5, 6]

Ingestion of SEB causes symptoms within 3 to 4 hours after exposure. These symptoms include severe weakness, abdominal cramps, nausea, vomiting, and explosive diarrhea.[1] Exposure to SEB aerosol can result in a robust systemic immune activation that resembled toxic shock.[3, 7, 8, 9, 10, 11] Mucosal exposure to SEB can also cause airway hyper-reactivity to resembles intrinsic asthma.[12] Symptoms from inhalational exposure to SEB include headache, fever, myalgia (muscle pain), scleral inflammation, leukocytosis, nausea, vomiting, and anorexia. Victims of exposure may suffer from chest pain and pulmonary edema.[1]

1. Ellison DH. Handbook of Chemical and biological Warfare Agents. Second Edition. © 2008 by Taylor & Francis group, LLC. Boca Raton. Page 483.

2. Lowell GH, Colleton C, Frost D, Kaminski RW, Hughes M, Hatch J, Hooper C, Estep J, Pitt L, Topper M, Hunt RE, Baker W, Baze WB. Immunogenicity and efficacy against lethal aerosol staphylococcal enterotoxin B challenge in monkeys by intramuscular and respiratory delivery of proteosome-toxoid vaccines. Infect Immun. 196 Nov;64(11):4686-93.

3. Rajagopalan G, Smart MK, Patel R, David CS. Acute systemic immune activation following conjunctival exposure to staphylococcal enterotoxin B. Infect Immun. 2006 oct;74(10)):6016-9.

4. Proft T, Fraser JD. 2003. Bacterial superantigens. Clin. Exp. Immunol. 133:299-306.

5. Burnett JC, Henchal EA, Schmaljohn AL, Bavari S. 2005. The evolving field of biodefense: therapeutic developments and diagnostics. Nat. Rev. Drug Discov. 4:281-297.

6. Madsen JM. 2001. Toxins as weapons of mass destruction: a comparison and contrast with biological-warfare and chemical-warfare agents. Clin. Lab. Med. 21:593-605.

7. Boles JW, Pitt ML, LeClaire RD, Gibbs PH, Ulricj RG, Bavari S. 2003. Correlation of body temperature with protection against staphylococcal enterotoxin B exposure and use in determining vaccine dose-schedule. Vaccine 21:2791-2796.

8. MattixME, hunt RE, Wilhelmsen CL, Johnson AJ, Baze WB. 1995. Aerosolized staphylococcal enterotoxin B-induced pulmonary lesions in rheus monkeys (Macaca mulatta). Toxicol. Pathol. 23:262-268.

9. Rajagopalan G, Sen M, Singh M, Murali N, Nath KA, Iijima K, Kita H, Leontovich AA, Unnikrishnan G, Patel R, David CS. 2006. Intranasal exposure to staphylococcal enterotoxin B elicits an acute systemic inflammatory response. Shock 25:647-656.

10. Roy CJ, Warfield KL, Welcher BC, Gonzales RF, Larsen T, Hanson J, David CS, Krakauer T, Bavari S. 2005. Human leukocyte antigen-DQ8 transgenic mice: a model to examine the toxicity of aerosolized staphylococcal enterotoxin B. Infect. Immun. 73:2452-2460.

11. Rajagopalan G, Sen MM, Singh M, Murali NS, Nath KA, Iijima K, Kita H, Leontovich AA, Gopinathan U, Patel R, David CS. Intranasal exposure to staphylococcal enterotoxin B elicits an acute systemic inflammatory response. Shock. 2006 Jun;25(6):647-56.

12. Herz U, Ruckert R, Wollenhaupt K, Tschernig T, Neuhaus-Steinmetz U, Pabst R, Renz H. Airway exposure to bacterial superantigen (SEB) induces lymphocyte-dependent airway inflammation associated with increased airway responsiveness—a model for non-allergic asthma. Eur J Immunol. 199 Mar;29(3):1021-31.

MYCOTOXINS

The wrecks of slavery are fast growing a fungus crop of sentiment.

William Dean Howells

D ISEASES CAUSED BY fungal or mold metabolites are known as mycotoxicoses. Molds make chemicals that are capable of causing disease or death in humans and animals. These chemicals known as mycotoxins are made by some species of molds in certain circumstances in order to protect their habitat from invasion by other micro-organisms. Some of the mycotoxins are beneficial such as penicillin and other antibiotics. However, other mycotoxins are potential chemical warfare agents. The mycotoxins that may be utilized as agents of bioterrorism include Aflatoxins, trichothecenes, and citrinin.[1] Mycotoxins can be used by even resource poor terrorist organizations to cause death or serious injury to human and animal health through poisoning of food and water sources. Humans and animals can suffer mycotoxin mediated toxicity through inhalation, ingestion, or skin and mucous membrane contact with a mycotoxin.[38] They may also be released in air borne form into confined crowded areas such as subways. Since molds capable of producing mycotoxins are easily grown of many common substrates including food stuffs, they provide terrorist will readily available bioterrorism agents.[36]

AFLATOXINS

THERE ARE FOUR major aflatoxins designated as B1, B2, G1, G2 with the B aflatoxins fluorescing blue under ultraviolet light and the G aflatoxins fluorescing green and the numeric designation of 1 or 2 referring to the relative rates of mobility during thin-layer chromatography.[1] Aflatoxins were first recognized after 100,000 turkey poults died in London in the early 1960s after consuming peanut meal that was contaminated with the mold *Aspergillus flavus*.[1, 2, 3] Aflatoxin B1 is the most potent naturally occurring cancer causing agent.[1, 4]

Aflatoxins are made by many strains of *Aspergillus flavus and Aspergillus parasiticus* with *Apsergillus flavus* being a common contaminate of agricultural products. However, aflatoxins may be produced by *Aspergilus nomius, Aspergillus bombycis, Aspergillus ochraceoroseus,* and *Aspergillus pseudotamari* although these species are less frequently occurring than *Aspergillus flavus* and *Aspergillus parasiticus*.[1, 5, 6, 7]

Aflatoxigenic molds are able to grow and produce aflatoxins in a broad range of substrates including cereals, oil-seed, tobacco, figs and nuts both in the field before harvest as well as during storage when there is sufficient moisture to promote mold growth.[1, 8, 9, 10, 11] Milk products can become contaminated with a hydroxylated form of aflatoxin B1 called aflatoxin M1 when cows consume aflatoxin contaminated feeds.[1, 12]

Aflatoxins are capable of causing cancer as well as non-cancerous mycotoxin related disease.[1, 13, 14] Food contaminated with aflatoxins have been associated with increased rates of liver cancer.[15] The International Agency for research on cancer has classified aflatoxin B1 as a human carcinogen.[16] Aflatoxin acts as a pro-carcinogen that is metabolized by the cytochrome P450 enzymes into a reactive 8,9-epoxide that caused cancer by binding to DNA and non-cancerous toxicity by binding to proteins.[1, 13]

Aflatoxin has been implicated as a possible agent of bioterrorism and as a biological warfare agent. There is substantial evidence that during the 1980s Iraq conducted research into the use of aflatoxin as a bio-warfare agent. The Iraqis were able to produce 2,300 liters of concentrated aflatoxin with a large portion of the aflatoxin concentrate being placed in missile warheads and the remainder being stockpiled.[1, 17, 18] The selection of a liver carcinogen by the Iraqis appears to have been at least in part motivated by the terror generated in populations by a known exposure to a potent human carcinogen.[19]

Diseases caused by consumption of aflatoxin are known as aflatoxicoses.[1] A 1974 episode of hepatitis in India after consumption of heavy aflatoxin contaminated maize resulted in the death of over 100 people. Some of the adults in this episode consume 2 to 6 mg of aflatoxin in a day.[1, 20] The acute lethal dose of aflatoxin has been estimated to be 10 to 20 mg.[1, 21] However, a woman who attempted to commit suicide by consuming over 40 mg of purified aflatoxin was alive and well 14 years later.[1, 22]

The carcinogenicity of aflatoxin is enhanced by concurrent infection with hepatitis B. The relative risk of liver cancer with aflatoxin exposure is 2. The relative risk of liver cancer with hepatitis B infection is 5. However, the risk of liver infection with combined exposure with aflatoxin exposure and hepatitis B infection is 60.[1, 23] It should be noted that the International Agency for Research on Cancer has classified aflatoxin B1 as a group I carcinogen.[1, 24]

CITRININ

C ITRININ IS A mycotoxin that may be produced by some *Penicillium* and *Aspergillus* species and was first isolated from *Penicillium citrinum* prior to World War II.[1, 25] Citrinin has also been isolated from *Monascus ruber* and *Monascus purpureus* which are industrial species of mold used in the production of red pigments.[1, 26] Citrinin is a nephrotoxin and has been found in wheat, rye, corn, oats, and rice.[1, 27, 28]

TRICHOTHECENES

TRICHOTHECENES ARE A family of mycotoxins that are produced by a number of commonly occurring molds including *Fusarium, Myrothecium, Trichothecium*, and *Stachybotrys*.[29] More than 200 trichothecenes have been reported.[29, 30] Trichothecenes are potent inhibitors of protein synthesis in eukaryotes.[29, 31] Trichothecene mycotoxins cause contamination of moldy grains and cereals resulting in diarrhea, vomiting, gastrointestinal hemorrhage, leukocytosis, shock and death.[29, 32] An outbreak of what is believed to have beentrichothene mycotoxin mediated disease occurred in 1944 in the area of Orenburg in Russia. This Siberia outbreak of animal and human disease was caused by consumption of moldy grain due to food shortages arising from World War II. Victims of this outbreak suffered from skin inflammation, vomiting, diarrhea, and hemorrhages. There was a mortality rate of over 10 percent.[33, 34] The specific trichothecene mycotoxin that is believed to have caused the Orenburg outbreak is T-2 toxin. Of the trichothecene mycotoxins, T-2 is considered to have the greatest potential as a bioterrorism agent.[35, 37] Chlorine dioxide either in solution or as a gas has been shown to effectively detoxify trichothecene mycotoxins.[39]

1. Bennett JW., Klich M. Mycotoxins. Clin Microbiol Rev. 2003 Jul;16(3): 497-516.
2. Blout WP. 1961. Turkey "X" disease. Turkeys 9:52, 55-58, 61, 77.
3. Goldblatt L. (ed). 1969. Aflatoxin, scientific background, control, and implications. Academic Press, New York, N.Y.
4. Squire RA. 1981. Ranking animal carcinogens: a proposed regulatory approach. Science 214:877-880.
5. Goto T, Wicklow DT, Ito Y. 1996. Aflatoxin and cyclopiazonic acid production by a sclerotium-producing *Aspergillus tamari* strain. Appl. Environ. Microbiol. 62:4036-4038.
6. Klich MA, Mullaney EJ, Daly CB, Cary JW. 2000. Molecular and physiological aspects of aflatoxin and sterigmatocystin biosynthesis by *A. tamari* and *A. ochraceoroseus*. Appl. Microbiol. Biotechnol. 53:605-609.
7. Peterson SW, Ito Y, Horn W, Goto T. 2001. *Aspergillus bombycis*, a new aflatoxigenic species and genetic variation in its sibling species, *A. nomius*. Mycologia 93:689-703.

8. Detroy RW, Lillehoj EB, Ciegler A. 1971. Aflatoxin and related compounds, p. 3-178. In A. Ciegler, S. Kadis, and S.J. Ajl (ed.), Microbial toxins, vol. VI: fungal toxins. Academic Press, New York, N.Y.

9. Diener UL, Cole RJ, Sanders TH, Payne GA, Lee LS, Klich MA. 1987. Epidemiology of aflatoxin formation by *Aspergillus flavus*. Annu. Rev. Phytopathol. 25:249-270.

10. Klich MA. 1987. Relation of plant water potential at flowering to subsequent cottonseed infection by *Aspergillus flavus*. Phytopathology 77:739-741.

11. Wilson DM, Payne GA. 1994. Factors affecting *Aspergillus flavus*group infection and aflatoxin contamination of crops, p. 309-325. In D.L. Eaton and J.D. Groopman (ed.). The toxicology of aflatoxins. Human health, veterinary and agricultural significance. Academic Press, San Diego, Calif.

12. Van Egmond HP. 1989. Aflatoxin M1: occurrence, toxicity, regulation, p. 11-55. In H.P. Van Egmond (ed.), Mycotoxins in dairy products. Elsevier Applied Science, London.

13. Eaton DL, Groopmen JD. 1994. The toxicology of aflatoxins: human health, veterinary, and agricultural significance. Academic Press, San Diego, calif.

14. Newberne PM, Butler WH. 1969. Acute and chronic effect of aflatoxin B1 on the liver of domestic and laboratory animals: a review. Cancer Res. 29:236-250.

15. Park S, Bae J, Nam BH, Yoo KY. Aetiology of cancer in Asia. Asian Pac J Cancer Prev. 2008 Jul-Sep;9(3):371-80.

16. Groopman JD, Johnson D, Kensler TW. Aflatoxin and hepatitis B virous biomarkers: a paradigm for complex environmental exposures and cancer risk. Cancer Biomark. 2005;1(1):5-14.

17. Stone R. 2001. Down to the wire on bioweapons talks. Science 293:414-416.

18. Zilinskas RA. 1997. Iraq's biological weapons. The past as future? J. Am. Med. Assoc. 276:418-424.

19. Stone R. 2002. Peering into the shadows: Iraq's bioweapons program. Science 297:1110-1112.

20. Krishnamachari KAVR, Bhat RV, Nagarajan V, Tilnak TMG. 1975. Hepatitis due to aflatoxicosis. An outbreak in Western India. Lancet i:1061-1063.

21. Pitt JI. 2000. Toxigenic fungi: which are important? Med. Mycol. 38(Suppl. 1):17-22.

22. Willis RM, Mulvihill JJ, Hoofnagle JH. 1980. Attempted suicide with purified aflatoxin. Lancet, i:1198-1199.

23. Ross RK, Yuan JM, Yu MC, Wogan GN, Qian GS, Tu JT, Groopman J, Gao YT, Henderson BE. 1992. Urinary aflatoxin biomarkers and risk of hepatocellular carcinoma. Lancet 339:1413-1414.

24. International Agency for research on cancer. 1982. The evaluation of the carcinogenic risk of chemicals to humans. IARC Monograph Supplement 4. International Agency for Research on Cancer, Lyon, France.

25. Hetherington AC, Raistrick H. 1931. Studies in the biochemistry of microorganisms. Part XIV. On the production and chemical constitution of a new yellow colouring matter, citrinin, produced from glucose by *Penicllium citrinum* Thom. Phil. Trans. R. Soc. London Ser. B 220B:269-295.

26. Blanc PJ, Loret MO, Goma G. 1995. Production of citrinin by various species of *Monascus*. Biotechnol. Lett. 17:291-294.

27. Carlton WW, Tuite J. 1977. Metabolites of *P. viridicatum* toxicology, p. 525-555. In J.V. Rodricks, C.W. Hesseltine, and M.A Mehlman (ed.), Mycotoxins in human and animal health. Pathotox Publications., inc., Park Forest South, Ill.

28. Abramson D, Usleber E, Marlbauer E. 2001. Immunochemical method for citrinin. P. 195-204. In M.W. Trucksess and A.F. Pohland (ed.), Mycotoxin protocols. Humana Press, Towoa, N.J.

29. Kimura M, Tokai T, Takahashi-Ando N, Ohsato S, Fujimura M. Molecular and genetic studies of fusarium trichothecene biosynthesis: pathways, genes, and evolution. Biosci Biotechnol Biochem. 2007 Sep;71(9):2105-23.

30. Grove JF. The Trichothecenes and their biosynthesis. Fortschr. Chem. Org. Naturst., 88, 63-130 (2007).

31. Ueno Y, Hosoya M, Ishikawa Y. Inhibitory effects of mycotoxins on the protein synthesis in rabbit reticulocytes. J. Biochem., 66, 419-422 (1969).

32. Pestka JJ, Smolinski AT. Deoxynivalenol: toxicology and potential effects on humans. J. Toxicol. Environ. Health B Crit. Rev., 8, 39-69 (2005).

33. Ueno Y. Trichothecenes: Overview address, in Rodericks JV, Hesseltine CW, Mehlman MA 9eds): Mycotoxins in Human and Animal Health. Park Forest South, IL: Pathtox, 1977, pp 189-208.

34. Casarett and Doull's Toxicology: The Basic Science of Poisons. 6th Edition. 2001. McGraw-Hill. New York. Page 1077.

35. Paterson RR. Fungi and fungal toxins as weapons. Mycol Res. 2006 Sep;110(Pt 9):1003-10.

36. Stark AA. Threat assessment of mycotoxins as weapons: molecular mechanisms of acute toxicity. J Food Prot. 2005 Jun;68(6):1285-93.

37. Henghold WB 2nd. Other biologic toxin bioweapons: ricin, staphylococcal enterotoxin B, and trichothecene mycotoxins. Dermatol clin. 2004 Jul;22(3):257-62, v.

38. Klassen-Fischer MK. Fungi as bioweapons. Clin Lab Med. 2006 Jun;26(2):387-95, ix.

39. Wilson SC, Brasel TL, Martin JM, Wu C, Andriychuk L, Douglas DR, Cobos L, Straus DC. Efficacy of chlorine dioxide as a gas and in solution in the inactivation of two trichothecene mycotoxins.

CATEGORY A BIOLOGICAL ORGANISMS

ANTHRAX

Yet will I bring one plague more upon Pharaoh, and upon Egypt.

Exodus

A NTHRAX IS A disease that in its dermatological manifestation is named after the Greek term for the black scabs that it produces.[3] Bacillus anthracis, the causative organism of anthrax was discovered by Robert Koch, the German microbiologist and physician, in 1877.[1, 2] Anthrax is an ancient disease that may have had its first reference in the bible as the fifth plague causing the death of cattle and the sixth plague in humans after the sprinkling of the ashes of cattle that had died from plague.[3] Following the 2001 terrorist attack on the World Trade Center in New York, there was a series of attacks where anthrax was sent through the mail to various public officials and members of the media. The first attack occurred on September 18, 2001 at the offices of American Media in Boca Raton, Florida. This attack resulted in the death of a journalist at America Media. The series of attacks ultimately caused the death of 5 persons.[6] It is estimated that a terrorist placing 100 kg of anthrax upwind of Washington D.C. would cause 130, 000 to 3 million deaths.[14] An incident involving release of aerosolized anthrax in 1979 from a microbiological weapons facility at Sverdlovsk in the Soviet Union resulted in at least 79 cases of anthrax infection and 68 deaths.[14, 16, 17] There appears to be general agreement among bioterrorism experts that the most likely candidate for a biological agent for a future terrorist attack is anthrax.[15]

Anthrax is a disease that occurs primarily in herbivorous animals including cattle, goats, sheep, pigs, and buffalo as well as elephants.[4] Animals become infected with anthrax through eating grass or soil that is contaminated with anthrax spores. Humans become infected with anthrax through contact with infected animals, or through the handling of hides or consumption of infected meat.[3] *Bacillus anthracis* is a soil organism that is distributed throughout the world.[4]

Anthrax has been considered as a biological weapon for many years.[22, 23, 24, 25] Attempts were make to weaponize anthrax for use against animals during World War I.[22, 26, 27] The spores of *Bacillus anthracis* are resistant to drying, heat, and chemicals. The British Army experimented with anthrax bombs during the Second World War. The British determined that the most effective way to use anthrax as a biological warfare agent is to disseminate an aerosol of a liquid suspension of bacteria in a bursting bomb.[22, 28, 29] N bombs which contained 106 special bomblets containing anthrax spores were dropped on Gruinard Island off the North West coast of Scotland. Gruinard Island was heavily contaminated with anthrax in 1941 due to anthrax bomb experimentation. The island remained contaminated for over 45 years and only became cleared of anthrax when the island was drenched with seawater and formaldehyde.[3, 22] The British also produced 5 million cattle cakes infected with anthrax to drop over Germany in order to infect the food supply of the Third Reich. However, World War II in Europe ended before the plan could be implemented.[22]

Bacillus anthracis is a gram positive rod shaped bacteria that forms spores. The spores of *Bacillus anthracis* are stable in water for up to 2 years and are resistant to the effects of chlorine for decades.[5]

There are three major forms of anthrax. The first form, cutaneous, occurs when spores are inoculated into minor skin abrasions. The second form, gastrointestinal, occurs when undercooked meat contaminated with anthrax is ingested. The third form, inhalational, occurs when the spores of anthrax are inhaled.[3] Of the three forms, inhalational anthrax is the most deadly. Inhalational studies conducted on monkeys during the 1950s indicated that approximately 10,000 spores were required to cause lethal disease in 50 percent of the animals. However, as few as one to three spores are capable of causing lethal disease if the spores reach the alveoli. Fine milling of anthrax spores can produce weaponized anthrax that is easily capable of reaching the alveoli if inhaled.[7]

Cutaneous anthrax initially presents as a small pimple that rises two to three days after anthrax spores have been inoculated into the skin. There is gradual enlargement of the pimple followed by vesicle formation. The vesicle then forms a blackish scab. The scab usually heals, however, in about one fifth of the cases there is bacteremia, meningitis, and death.[8] In addition to the black eschar (scab), there may be massive soft tissue edema formation in cutaneous anthrax.[14] The most common areas of the body involved with cutaneous anthrax are the head, neck and extremities.[18] Cutaneous anthrax can cause bacteremia and toxemia with a fatality rate of 20 to 25 percent in untreated causes.[18, 19, 20, 21]

Gastrointestinal anthrax may cause oropharyngeal edema with accompanying respiratory compromise necessitating surgical airway placement. There may also be intestinal ulceration, necrosis, bleeding, and perforation requiring surgical exploration and resection of affected segments of the gastrointestinal tract.[14]

Inhalational anthrax has an initial presentation similar to influenza. However, after 3 to 4 days of influenza symptoms, victims of inhalational anthrax develop severe shortness of breath accompanied by either fever or hypothermia. The mortality rate from inhalational anthrax is 80 to 90 percent.[8] Inhalational anthrax can result in systemic sepsis and massive ascites, derangements of electrolytes, and profound shock necessitating aggressive resuscitation of fluids and careful hemodynamic monitoring.[14]

Children with anthrax present with signs and symptoms which may differ from those in adults. The clinical presentation of anthrax in children is varied with a high mortality rate for gastrointestinal and inhalational anthrax as well as anthrax meningoencephalitis. Children with gastrointestinal anthrax present with upper gastrointestinal tract disease characterized with dysphagia and oropharyngeal findings and with lower tract disease manifested by abdominal pain, fever, nausea, and vomiting. Children with inhalational disease present with atypical finding that primarily include meningoencephalitis. The differences in presentation of anthrax in adults and in children present a need for additional preparedness by primary care and emergency physicians in the early recognition of an outbreak.[10, 11]

The treatment of choice for anthrax remains penicillin. In a recent review of anthrax cases in Turkey, the mortality rate for persons receiving penicillin was 2.8 percent.[9] Fulminate anthrax has a high mortality despite modern advances in supportive medical care. Initiation of prophylactic treatment with antibiotics and anthrax antiserum during the prodromal phase is associated with significantly improved survival.[12] However, survivors of inhalational anthrax tend to report significantly lower overall physical heath than survivors of cutaneous anthrax.[13]

1. McNeill WH. Plagues and Peoples. 1976. Anchor Press. Garden City, New York. Page 236.

2. Watts S. Epidemics and History: Disease, Power and Imperialism. 1997. Yale University Press. New Haven, Connecticut. Page 4.

3. Hart T. Microterrors: the complete guide to bacteria, viral and fungal infections that threaten our health. © 2004 by Axis Publishing Ltd. Buffalo, New York. Page 100.

4. Goldman L, Bennett JC. Cecil Textbook of Medicine. 21st Edition. 2000. W.B. Saunders Company. Philadelphia. Page 1704.

5. Ellison DH. Handbook of Chemical and biological Warfare Agents. Second Edition. © 2008 by Taylor & Francis group, LLC. Boca Raton. Page 498.

6. Burke R. Counter-Terrorism for Emergency responders. Second Edition. © 2007 by Taylor & Francis group, LLC. Boca Raton. Page 50.

7. Harrison's Principles of Internal Medicine. 16th Edition. © 2005 by the McGraw-Hill Companies. New York. Page 1280.

8. Hart T. Microterrors: the complete guide to bacteria, viral and fungal infections that threaten our health. © 2004 by Axis Publishing Ltd. Buffalo, New York. Page 101.

9. Doganay M, Metan G. Human Anthrax in Turkey from 1990 to 2007. Vector Borne Zoonotic Dis. 2008 Oct 22.

10. Bravata DM, Wang E, Holty JE, Lewis R, Wise PH, Nayak S, Liu H, McDonald KM, Owens DK. Pediatric anthrax: implications for bioterrorism preparedness. Evid Rep Technol Assess. 2006 Aug;(141):1-48.

11. Bravata DM, Holty JE, Wang E, Lewis R, Wise PH, McDonald KM, Owens DK. Inhalational, gastrointestinal, and cutaneous anthrax in children: a systemic review of cases: 1900 to 2005. Arch Pediatr Adolesc Med. 2007 Sep;161(9):896-905.

12. Holty JE, Bravata DM, Liu H, Olshen RA, McDonald KM, Owens DK. Systemic review: a century of inhalational anthrax cases from 1900 to 2005. Ann Intern Med. 2006 Feb 21;144(4): 270-80.

13. Reissman DB, Whitney EA, Taylor TH Jr, Hayslett JA, Dull PM, Arias I, Ashford DA, Bresnitz EA, Tan C, Rosenstein N, Perkins BA. One-year health assessment of adult survivors of Bacillus anthracis infection. JAMA. 2004 Apr 28;291(16):1994-8.

14. Binkley CE, Cinti S, Simeone DM, Colletti LM. Bacillus anthracis as an agent of bioterrorism: a review emphasizing surgical treatment. Ann Surg. 2002 Jul;236(1):9-16.

15. Zhou B, Wirsching P, Janda KD. Human antibodies against spores of the genus bacillus: a model study for detection of and protection against anthrax and the bioterrorist treat. Proc Natl Acad Sci USA. 2002 Apr 16;99(8):5241-6.

16. Meselson M, Guillemin J, Hughes-Jones M, et al. The Sverdlovsk anthrax outbreak of 1979. Science 1994;266:1202-1208.

17. Abramova FA, Grinberg LM, Yampolskaya OV, et al. Pathology of inhalational anthrax in 42 cases from the Sverdlovsk outbreak of 1979. Pro Natl Acad Sci USA 1993; 90:2291-2294.

18. Shieh WJ, Guarner J, Paddock C, Greer P, Tatti K, Fischer M, Layton M, Philips M, Bresnitz E, Quinn CP, Popovic T, Perkins BA, Zaki SR, Anthrax Bioterrorism Investigation Team. The critical role of pathology in the investigation of bioterrorism-related cutaneous anthrax. Am J Pathol. 2003 Nov;163(5):1901-10.

19. LaForce FM. Anthrax. Clin Infect Dis. 1994. 19:1009-1013.

20. Gold H: Anthrax: a report of 117 cases. Arch Int Med 1955. 96:387.

21. McSwiggan DA, Hussain KK, Taylor IO: A fatal case of cutaneous anthrax. J Hyg (Lond) 1974. 73:151-156.

22. Spencer RC. Bacillus anthracis. J Clin Pathol. 2003 Mar;56(3):182-7.

23. Spencer RC. Wilcox MH. Agensts of biological warfare. Rev Med Microbiol. 1993;4:138-43.

24. Inglesby TV, Henderson DA, Bartlett JG, et al. Anthrax as a biological weapon. JAMA 1999;281:1735-45.

25. Specncer RC, Lightfoot NF. Preparedness and response to bioterrorism. J Infect. 2001;43:104-10.

26. Redmond C, Pearce MJ, Manchee RJ, et al. Deadly relic of the Great War. Nature 1998:393:747-8.

27. Wheelis M. Biological sabotage in World War I. In: Geissler E, van Courtland Moon JE, eds. Biological and toxin weapons: research, development and use from the Middle Ages to 1945. Oxford: Oxford University Press, 1999:35-62.

28. Carter GB. Chemical and biological defense at Porton Down: 1916-2000. London: HMSO, 2000.

29. Carter GB. Biological warfare and defense in the United Kingdom: 1940-1979. Journal of the Royal United Services Institute for Defense Studies 1992;137:67-74.

PLAGUE

From winter, plague and pestilence, good Lord, deliver us!

Thomas Nashe

P LAGUE IS CAUSED by *Yersinia pestis* which is a gram negative oval shaped facultative anaerobic bacterium that is non-motile and does not form spores.[1] Alexandre Yersin discovered the plague bacillus in 1894.[10] Plague has a clinical presentation of sudden onset of fever, chills, body aches, weakness, headaches, vomiting, and nausea after an incubation period of 3 to 7 days. There are three forms of plague depending upon the route of infection. The bubonic form which results from a bite from an infected flea is the most common. The pneumonia form has human to human transmission through inhalation of infected respiratory droplets. The third form, septicaemic, is an intermediary form of plague leading from bubonic to pneumonic plague.[10] *Yersinia pestis* causes several thousand cases of human disease annually.[10, 11, 12] *Yersinia pestis* is transmitted by flea bites and causes a rapidly progressive disease in its bubonic form that has a high mortality rate if untreated. Persons with bubonic plague may develop pneumonia due to plague bacillus spread through the blood. This pneumonic plague is virtually always fatal and can be transmitted from person to person by airborne droplet spread.[10]

Besides *Yersinia pestis*, human disease may be caused by *Yersinia enterocolitica* and *Yersinia pseudotuberculosis*. These two organisms are transmitted by oral-faecal spread and cause moderately intense intestinal symptoms.[10]

It is believed that *Yersinia pestis* emerged during the last 1,500 to 20,000 years as a clone of *Yersinia pseudotuberculosis*.[10, 13, 14] During the course of its evolution from *Yersinia pseudotuberculosis*, two plasmids were acquired by *Yersinia pestis* that allowed flea-borne transmission of infection.[10, 15, 16] This evolution explains the great pathogenicity of *Yersinia pestis* organism compared to *Yersinia enterocolitica* and *Yersinia pseudotuberculosis* since septicemia must occur in the victim in order for a transmitting flea to acquire the infection through a blood meal.[10, 17]

Yersinia pestis can survive in water for up to 16 days. At temperature near or around freezing, the organism can survive for months and even years. However, *Yersinia pestis* is destroyed by several hours of exposure to sunlight.[1]

Plague is believed to have begun as an infection of gerbils in eastern Asia.[2] Plague has occurred in three pandemics. The first pandemic of plague spread around the Mediterranean Sea starting in 542 A.D. during the reign of the Eastern Roman emperor Justinian.[10] This pandemic probably occurred due to the spread of the plague from its endemic homeland in the area of the Himalayan frontiers of India and China.

[7, 8, 9] The second pandemic struck Europe in 1348 after a Mongol army catapulted the bodies of plague victims into the Crimean trading outpost of the Genoese at Kaffa (Feodosiya, Ukraine) in the Crimean.[7, 8, 9] This second pandemic is believed to have killed one quarter to one third of the population of Europe and continued to recur intermittently for 300 years.[10] The third pandemic began in Yunan, China in 1860 and lead to the spread of plague to the Western United States.[3, 4] Each of pandemics was thought to have been caused by a different variant of *Yersinia pestis*. The first pandemic is thought to have been caused by the Antiqua biovar. Antiqua still exists in Central Asia and Africa. The second pandemic was caused by Medievalis which is currently existent in central Asia. The third pandemic was caused by Orientalis which currently has an almost world wide distribution.[10, 18]

The second pandemic which is also known as the Black Death had a devastating impact upon Medieval Europe with major influence upon the art, culture, politics, religion, and socio-economic development of the continent.[10, 19, 20] The long held belief that the Black Death was caused by *Yersinia pestis* has recently been questioned.[10, 21, 22, 23] However, the discovery of genetic material from *Yersinia pestis* in the graves of medieval victims of the Black Death is supportive of the theory that the Black Death was caused by *Yersinia pestis*.[10, 24]

During World War II Unit 731 of the Japanese army is believed to have dropped fleas infected with plague over Manchuria. The United States and the Soviet Union each conducted research concerning the use of *Yersinia pestis* as a bio-warfare agent.[5, 10, 27, 28]

There are two forms of plague. Bubonic plague presents as acute swelling of regional lymph nodes. The swollen lymph nodes are called "bubos". Untreated bubonic plague has a mortality of approximately 50 percent. The other form of plague is pneumonic plague. Pneumonic plague is an infection of the lungs by *Yersinia pestis*. The mortality of untreated pneumonic plague is virtually 100 percent.[6] Human to human spread of plague may occur through pneumonic transmission of respiratory droplets or in the bubonic form through flea bites.[10, 25, 26]

The potential for use of plague by terrorists is high. If terrorists can weaponize the plague bacillus so that it can be released in aerosolized form into confined areas such as buildings and airplanes, the fear generated by the reputation of the agent of the Black Death would be great.[10, 27, 28]

1. Ellison DH. Handbook of Chemical and Biological Warfare Agents. Second Edition. © 2008 by Taylor & Francis group, LLC. Boca Raton. Page 520.
2. McKeown T. The Origins of Human Disease. 1988. Basil Blackwell Ltd. Oxford, England. Page 56.
3. Hart T. Microterrors: the complete guide to bacteria, viral and fungal infections that threaten our health. © 2004 by Axis Publishing Ltd. Buffalo, New York. Page 140.

4. Harrison's Principles of Internal Medicine. 16th Edition. © 2005 by the McGraw-Hill Companies. New York. Page 1282.

5. Harrison's Principles of Internal Medicine. 16th Edition. © 2005 by the McGraw-Hill Companies. New York. Page 1283.

6. Goldman L, Bennett JC. Cecil Textbook of Medicine. 21st Edition. 2000. W.B. Saunders Company. Philadelphia. Page 1701.

7. 7. Saikaly PE, Barlaz MA, de Los Reyes FL 3rd. Development of quantitative real-time PCR assays for detection and quantification of surrogate biological warfare agents in building debris and leachate. Appl Environ Microbiol. 2007 oct;73(20):6557-65.

8. Lim DV, simpson JC, Kearns EA, Kramer MF. 2005. Current and developing technologies for monitoring agents of bioterrorism and biowarfare. Clin. Microbiol. Rev. 18:583-607.

9. Szinicz L. 2005. History of chemical and biological warfare agents. Toxicology 214:161-181.

10. Stenseth NC, Atshabar BB, Begon M, Belmain SR, Bertherat E, Carniel E, Gage KL, Leirs H, Rahalison L. Plague: past, present, and future. PLoS Med. 2008 Jan 15;5(1):e3.

11. World Health Organization. 2003. Plague. Wkly Epidemiol Rec 78, 253-260.

12. World Health Organization. 2005. Plague. Wkly Epidemiol Rec 80 138-140.

13. Achtman M, Zurth K, Morrelli G, Torrea G, Guiyoule A, et al. 1999. *Yersinia pestis*, the cause of plague, is a recently emerged clone of *Yersinia pseudotuberculosis*. Proc Natl Acad Sci USA 96: 14043-14048.

14. Achtman M, Morelli G, Zhu P, Wirth T, Diehl I, et al. 2004. Microevolution and history of the plague bacillus, *Yersinia pestis*. Proc Natl Acad Sci USA 101: 17837-17842.

15. Sodeinde OA, Subrahmanyam YV, Stark K, Quan T, Bao Y, et al. 1992. A surface protease and the invasive character of plague. Science 258: 1004-1007.

16. Hinnebusch BJ, Rudolph AE, Cherepanov P, Dixon JE, Schwan TG, et al. 2002. Role of Yersinia murine toxin in survival of *Y. pestis* in the midgut of the flea vector. Science 296: 733-735.

17. Carniel E. 2003. Evolution of pathogenic *Yersinia*, some lights in the dark. Adv Exp Med Biol 529:3-12.

18. Guiyoule A, Grimont F, Iteman I, Grimont PAD, Lefevre M, et al. 1994. Plague pandemics investigated by ribotyping of *Yersinia pestis* strains. J Clin Microbiol 32: 634-641.

19. Twigg G. 1984. The Black Death: a biological reappraisal. London: Batsform Academic and Educational.

20. Ziegler P. 1969. The Black Death. Wolfeboro Falls (NH): Alan Sutton Publishing.
21. Scott S, Duncan CJ. 2001. Biology of plagues: evidence from historical populations. Cambridge: Cambridge University Press.
22. Cohn SK Jr. 2002. The Black Death transformed. London: Arnold.
23. Drancourt M. 2006. *Yersinia pestis* as a telluric, human ectoparasite-borne organism. Lancet Infect Dis 6: 234-241.
24. Raoult D, Aboudharam G, Crubezy E, Larrouy G, Ludes B, et al. 2000. Molecular identification by "suicide PCR" of *Yersinia pestis* as the agent of Medieval Black Death. Proc Natl Acad Sci USA 97: 12800-12803.
25. Pollitzer R. 1954. Plague. Monogr Ser World Health Organ 22: 1-698.
26. Pollitzer R. 1960. A review of recent literature on plague. Bull World Health Organ 23: 313-400.
27. Inglesby TV, Dennis DT, Henderson DA, Bartlett JG, Ascher MS, et al. 2000. Plague as a biological weapon: medical and public health management. JAMA 283: 2281-2290.
28. Koirala J. 2006. Plague: disease, management, and recognition of act of terrorism. Infect Dis Clin North Am 20: 273-287.

SMALLPOX

A propos of distempers, I am going to tell you a thing that I am sure will make you wish your selfe here. The Small Pox so fatal and so general amongst us is here entirely harmless by the invention of engrapfing. There is a set of old Women who make it their business to perform the Operation.

Lady Mary Wortley Montagu

SMALLPOX IS AN ancient disease caused by the *Variola* virus which is a member of the *Poxviridae* family of DNA viruses.[10] The *Poxviridae* (poxviruses) family of viruses replicate in the cytoplasm of host cells and have large DNA genomes that encode about 200 proteins. The *Poxviridae* family is subdivided into two subfamilies consisting of the *Chordopoxvirinae* and the *Entomopxvirinae* subfamilies. The *Chordopoxvirinae* are further subdivided into eight genera consisting of *Avipoxvirus, capripoxvirus, Molluscipoxvirus, Suipoxvirus, Leporipoxvirus, Yatapoxvirus, Parapoxvirus,* and *Orthopoxvirus.*[10, 11] *Variola* virus (VARV), the virus causing smallpox disease in humans, is a member of the *Orthopoxvirus* genus. VARV is futher divided into two phenotypic subtypes based upon the case fatality due to infection. The variola major subtype has a case fatality rate of up to 40 percent in unvaccinated persons. The variola minor subtype has a case fatality rate of approximately 1 percent in unvaccinated persons.[10, 12]

Smallpox is estimated to have killed more persons throughout history than all other infectious disease combined.[10, 11] The mummy of Pharaoh Ramses V who died in 1157 B.C. had scars from smallpox.[1] Smallpox was brought to Europe by Moorish invaders. The Spanish conquistadors brought smallpox to the Americas in the 16th century. Before the inception of vaccination, between 7 to 12 percent of all deaths were due to smallpox.[2]

On May 8, 1980 the Thirty-third World Health Assembly declared smallpox to be eradicated.[16] Since that time the only known samples of smallpox virus have been stored in repositories at the Centers for Disease Control and Prevention in Atlanta, Georgia and the State Research Center of Virology and Biotechnology at Koltsovo in the Novosibirsk region of Russia.[10] Since humans are the only reservoir and vector of small pox, the only known samples of smallpox are in the American and Russia repositories.[3, 4] Since vaccination for smallpox was discontinued in the United States in 1972, almost 50 percent of the population in the United States is fully susceptible to smallpox.[4]

50

In the event that there is an outbreak of smallpox due to a bioterrorist attack, victims suffering from smallpox will tend to be concentrated into healthcare facilities. There is increasing evidence that aerial spread of pathogens can be a serious risk factor for nosocomial infections in hospitals and other healthcare facilities.[9] Spread of smallpox (variola major) through airborne spread of droplet nuclei is a potential means of the spread of the disease from person to person.[5, 6, 7] It has been shown that placing ultraviolet lights in the ceilings of healthcare facilities including emergency departments and waiting rooms can decrease viable viral aerosols including the smallpox virus.[5] Use of ultraviolet germicidal irradiation (UVGI) may be an effective means of decreasing viral spread in locations where crowds congregate.[5, 8]

Mathematical models of epidemic spread are widely used in planning public health responses to bioterrorist attacks.[13, 14] There is a lack of certainty concerning the accuracy of our epidemic models concerning the spread of smallpox in the event of a bioterrorist attack.[13] However, there is evidence based upon epidemic modeling that regional mass vaccination for smallpox can be effective in halting the spread of the disease.[15] Consequently, it is important that any outbreak of smallpox be rapidly identified so that effective quarantine and fomite decontamination activities can occur before the disease spreads to the general population. Polymerase Chain Reaction (PCR) can be utilized to rapidly identify the smallpox virus.[10]

1. Burke R. Counter-Terrorism for Emergency responders. Second Edition. © 2007 by Taylor & Francis group, LLC. Boca Raton. Page 158.

2. Hart T. Microterrors: the complete guide to bacteria, viral and fungal infections that threaten our health. © 2004 by Axis Publishing Ltd. Buffalo, New York. Page 78.

3. Ellison DH. Handbook of Chemical and biological Warfare Agents. Second Edition. © 2008 by Taylor & Francis group, LLC. Boca Raton. Page 578.

4. Harrison's Principles of Internal Medicine. 16th Edition. © 2005 by the McGraw-Hill Companies. New York. Page 1284.

5. McDevit JJ, Milton DK, Rudnick SN, First MW. Inactivationof poxviruses by upper-room UVC light in a simulated hospital room environment. PLoS ONE. 2008 Sep 10;3(9):e3186.

6. Wehrle PF, Posch J, Richter KH, Henderson DA (1970) An Airborne outbreak of smallpox in a German hospital and its significance with respect to other recent outbreaks in Europe. Bull World Health Organ 43: 669-79.

7. Henderson DA, Ingleby TV, Bartlett JG, Ascher MS, Eitzen E, et al. (1999) Smallpox as a biological weapon: medical and public health management. Working Group on Civilian Biodefense. Jama 281:2127-37.

8. Brickner PW, Vincent RL, First M, Nardell E, Murray M, Kaufman W. The application of ultraviolet germicidal irradiation to control transmission of airborne disease: bioterrorism countermeasure. Public Health rep. 2003 Mar-Apr;118(2):99-114.

9. Beggs CB, Kerr KG, Noakes CJ, Hathway EA, Sleigh PA. The ventilation of multiple-bed hospital wards: review and analysis. Am J Infect Control. 2008 May;36(4):250-9.

10. Sulaiman IM, Tang K, Osborne J, Sammons S, Wohlhueter RM. J Clin Microbiol. 2007 Feb;45(2):358-63.

11. Moss B. 1996. Genetically engineered poxviruses for recombinant gene expression, vaccination, and safety. Proc. Natl. Acad. Sci. USA 93:11341-11348.

12. Massung RF, Liu L, QI J, Knight JC, Yuran TE, Kerlavage AR, Parson JM, Venter JC, Esposito JJ. 1994. Analysis of the complete genome of smallpox *Variola* major strain Bangladesh-1975. Virology 201:215-240.

13. Elded BD, Dukic VM, Dwyer G. Uncertainty in predictions of disease spread and public health responses to bioterrorism and emerging diseases. Proc Natl Acad Sci USA. 2006 Oct 17;103(42):15693-7.

14. Ferguson NM, Keeling MJ, Edmunds WJ, Gant R, Grenfell BT, Anderson RM, Leach S. 2003. Nature 425:681-685.

15. Riley S, Ferguson NM. Smallpox transmission and control: spatial dynamics in Great Britain. Proc Natl Acad Sci USA. 2006 Aug 15;103(33); 12637-42.

16. Pennington H. Smallpox and bioterrorism. Bull World Health Organ. 2003; 81(10):762-7.

CHEMICAL AGENTS OF TERROR

DULCE ET DECORUM EST

Bent double, like old beggars under sacks,
Knock-kneed, coughing like hags, we cursed through sludge,
Till on the haunting flares we turned our backs,
And towards our distant rest began to trudge.
Men marched asleep. Many had lost their boots,
But limped on, blood-shod. All went lame, all blind;
Drunk with fatigue; deaf even to the hoots
Of gas-shells dropping softly behind.

Gas! Gas! Quick, boys! An ecstasy of fumbling,
Fitting the clumsy helmets just in time,
But someone still was yelling out and stumbling
And floundering like a man in fire or lime.
Dim through the misty panes and thick green light,
As under a green sea, I saw him drowning.
In all my dreams, before my helpless sight,
He plunges at me, guttering, choking, drowning.

If in some smothering dreams, you too could pace
Behind the wagon that we flung him in,
And watch the white eyes writhing in his face,
His hanging face, like a devil's sick of sin;
If you could hear, at every jolt, the blood
Come gargling from the froth-corrupted lungs,
Obscene as cancer, bitter as the cud
Of vile, incurable sore on innocent tongues,
My friend, you would not tell with such high zest
To children ardent for some desperate glory,
The old Lie: Dulce et decorum est Pro patria mori.

Wilfred Owen
Poet of the First World War

CHEMICAL WARFARE AGENTS have a long history. The ancient Greek states of Athens and Sparta in 423 B.C. used the poisonous gases emanating from burning sulfur, pitch, and wax in their war.[1] Chemical weapons were proposed by both sides during the American Civil War. Agents that were suggested included black and cayenne pepper, snuff, and mustard, hydrogen cyanide, chloroform, chlorine, arsenic compounds, sulfur, and acids. The most common proposed delivery device was explosive artillery projectiles. None of the proposed chemical weapons were actually used during the American Civil War.[1, 2] The 1899 Treaty of the Hague prohibited the use of ammunition containing poisonous gases.[1] However, despite the Treaty of the Hague, the Germans conducted the first large scale chemical warfare attack with chlorine gas during World War I on April 22, 1915.[3, 4, 5, 6, 7] Fritz Haber, a Nobel laureate who developed a practical process to create nitrogen fertilizer from air, was at the forefront of the German chemical warfare program.[8, 9, 10] The G-Series nerve agents tabun, sarin, and soman were developed in Germany during the 1930s. During the mid-1930s, the Italians used phosgene and mustard gas in its war with Ethiopia. In 1938, the Japanese used mustard gas when they invaded China.[5] Although Nazi Germany stockpiled tabin and sarin during the Second World War, these agents were not used.[11, 12] During and following the Second World War, the Allied powers and the Soviet Union developed large stockpiles of chemical warfare agents. These stockpiles remain in various stages of decay including large numbers of chemical bombs abandoned in uncharted dumps.[5, 13] Iraq and Iran both used chemical weapons in war from 1980 to 1988. The Iraq government used chemical weapons against the Kurds in 1985 with a fatality rate of 60 percent.[5, 14, 23] The Japanese cult, Aum Shinriko used sarin in a terrorist attack in the Tokyo subway in 1995.[1, 23]

Chemical terrorism presents a danger because the widespread use of industrial chemicals closed to areas of high population density provides readily made supplies of weapons to terrorists.[21, 22] Chemical warfare agents are generally divided into four categories; choking agents, blood agents, blistering agents, and nerve agents.

CHOKING AGENTS

C HOKING CHEMICAL AGENTS cause injury and death by their effects on the lungs. The choking agents literally cause exposed persons to choke. Two of the prototypical choking agents are chlorine and phosgene gas.

Chlorine is a yellowish-green gas at room temperature that has a pungent odor and is strongly irritating to mucous membranes. Chlorine has many industrial uses including production of chlorinated organic polymers, solvents, and other organic chemicals. Since it is denser than air, chlorine gas tends to collect in depressions in the ground.[3, 15] The greatest mass exposures to chlorine occurred as the result of use of chlorine gas during World War I.[18]

Chlorine species are highly reactive with tissue injury occurring with exposure to chlorine, hydrochloric acid, hypochlorous acid, and chloramines.[18] the toxicology of chlorine s related almost entirely to its effects on the respiratory system.[24] An acute exposure to chlorine causes mucous membrane irritation, cough, shortness of breath, chest tightness, and hemoptysis. Exposure victims may present with wheezing, hypoxia, and tachycardia.[3, 15, 16] The effects of chlorine exposure are dose related with the magnitude of tissue damage being greater with greater levels of exposure. Pulmonary function testing conducted immediately after exposure may show either obstructive or restrictive findings that resolve over time in the majority of cases. However, high levels of exposure may result in permanent abnormalities on pulmonary function testing.[18] Since chlorine is used widely as an industrial chemical, individuals can be exposed to chlorine through industrial and transportation accidents as well as misuse of domestic cleaners. Although must persons with chlorine exposure recover normal pulmonary function, exposure to chlorine can cause a variety of lung injuries including pulmonary edema, restrictive lung disease including pulmonary fibrosis, and obstructive lung diseases such as Reactive Airways Dysfunction Syndrome.[17] Exposure victims may also suffer bronchiolitis obliteran.[19] Certain subpopulations such as smokers and persons with atopy may be at increased risk of lung injury with chlorine exposure.[17, 18, 20] Treatment of chlorine exposures is mainly supportive.[24] Treatment for severe exposures to chlorine often includes the use of corticosteroids, however, the efficacy of this treatment is uncertain.[19, 24]

Phosgene is a colorless gas at normal atmosphere pressure and room temperature with a strong odor of fresh mowed hay. Phosgene was first synthesized in 1812 by Humphrey Davis. Fritz Haber prepared it for use as a chemical weapon for the

Germans. It was first used in warfare on December 19, 1915 and resulted in 1069 casualties and 120 deaths. The Allies also used phosgene later in the war.[3, 25]

Phosgene is used in a wide variety of industrial application including the manufacture of dyes, resins, foams, and polymers. Consequently, it is readily available as a potential chemical terrorism agent.[3]

Exposure to phosgene causes immediate coughing and choking, tearing of the eyes, headache, and chest tightness. In addition, victims of exposure can suffer from nausea and vomiting. These initial symptoms are usually followed by a 2 to 24 hour period of when the victim feels well and symptom free. This period is followed by symptoms of coughing, shortness of breath, rapids breathing, and cyanosis due to pulmonary edema from increased alveolar pulmonary capillary permeability.[3, 26] No antidote exists for phosgene exposure and care is supportive. While steroids are frequently used, the use of steroids remains unproven.[3, 25, 26]

BLOOD AGENTS

B LOOD AGENTS DESTROY the ability of blood to effectively provide oxygen to the tissues of the body. Cyanide is the classic blood agent. Hydrogen cyanide is a colorless liquid with vapors that are lighter than air. Cyanide can be absorbed in both its liquid and gaseous form through the skin and through the gastrointestinal tract.[27] Since it is lighter than air, cyanide in gaseous form is easily dispersed in the open air. This caused cyanide to be ineffective as a warfare agent during World War I. The Nazis used hydrogen cyanide in the form of Zyklon B in gas chambers. [27, 28] Cyanide causes cellular hypoxia by disrupting electron transport through the cytochrome a-cytochrome a-3 system. Due the inability of the body to utilize oxygen at the tissue level, venous blood has a high oxygen concentration in cases of cyanide poisoning. The same mechanism of toxicity occurs with hydrogen sulfide. Death due to cyanide poisoning occurs within seconds or minutes of high concentrations of cyanide or sulfide gas as a result of central respiratory arrest.[29]

BLISTERING AGENT

B LISTERING OR VESICANT agents cause cutaneous blisters, damage to the
respiratory tract, lesions of the eye, and depression of the bone marrow.[30] Tons of
mustard gas were made for use as a chemical warfare agent. Much of these stockpiles
were buried in landfills, disposed at sea, and left in a decaying state in storage facilities.[32]
Sulphur mustard was heavily used during World War I and caused over 100,000
casualties during the Iran-Iraq War.[31] Mustard agents can cause chronic obstructive
lung disease, pulmonary fibrosis, recurrent corneal ulcers, chronic conjunctivitis,
abnormal pigmentation of the skin, and cancer. Despite extensive research during
the last 90 years, there remains no specific antidote to mustard agents.[30] However,
hypothermia has been proposed as an adjunct to therapy for vesicant induced skin
injuries.[39]

Sulphur mustard was first synthesized by Despretz in 1833 and modified by
Niemann and Guthrie in 1860.[38] Sulphur mustard is easy and cheap to synthesize
making it a potential chemical terrorism agent. Sulphur mustard rapidly acts as an
alkylating agent that causes disruption of nucleic acids and proteins resulting in
impairment of cell homeostasis and cell death.[31]

Victims of mustard agent exposures have immediate symptoms involving the eyes,
skin, and respiratory tract and delayed effects upon the nervous, cardiac, and digestive
system. Symptoms include malaise, anorexia, lacrimation, salivation, respiratory
distress, vomiting, and hyper-excitability. High doses can cause necrosis of the skin
and mucous membranes including the mucous membranes of the respiratory tract,
bronchitis, bronchopneumonia, hemo-concentration, leucopenia, intestinal lesions,
convulsions, and death.[33] The term mustard gas is a misnomer since the mustard
chemicals are liquid at room temperature. There is a garlic odor to sulfur mustard
agents.[34] Sulfur mustards may cause defective spermatogenesis persisting for many
years after exposure.[41, 42]

Lewisite [dichloro(2-chlorovinyl) arsine] is not only a vesicant but also a systemic
poison with liver, gallbladder, and renal and urinary tract toxicity. Lewisite has
approximately the same inhalational toxicity as mustard agent. However, lewisite is
more rapidly acting and can cause immediate death due to "lewisite shock" which
is a loss of blood plasma resulting from increased permeability of capillaries due to
circulating Lewisite.[34, 35, 36, 37] Lewisite was developed as an arsenical vesicant as a
counter to German war gas agents during the First World War.[39,]

NERVE AGENTS

THE ORGANOPHOSPHORUS NERVE warfare agents soman, sarin, VX, and tabun are among the most deadly man-made chemicals.[43, 44] Sarin was used against Kurdish citizens of Halabja in 1988 by the Iraqi regimen with a death count of approximately 5,000 persons.[43,45] The Japanese cult Aum Shinriko used sarin in a 1995 terrorist attack on the Tokyo subway that resulted in the death of 12 persons and made nearly one thousand ill.[43, 46] Organophosphorus nerve agents were developed in Germany during the Third Reich. The development of these nerve agents occurred due to a collaboration between the Nazi administration, industry, and academia. The first nerve agent, tabun, was synthezied by chance during research on organophosphate pesticides by Gerald Schrader, an I.G. Farben company. The neuro-toxicity of tabun was discovered by the accidental exposure of a research associate to the agent. The discovery of tabun lead to the synthesis of sarin and soman as potential warfare agents.[50]

The LD50 for percutaneous exposure to organophosphorus nerve warfare agents ranged from 10 mg/person for VX to 1000 mg/person for tabun.[43, 44] Organophosphorus nerve agents primarily function as irreversible acetylcholinesterase inhibitors causing a rapidly progressive cholinergic crisis.[47, 48] The inhibition of the enzyme acetylcholine esterase by organophosphorus nerve agents cause the synaptic build-up of acetylcholine. This resulted in profound over stimulation of muscarinic and nicotinic receptors in the autonomic and central nervous system. Parasympathetic over-stimulation produces sialorhea, miosis, bronchospasm, and bronchorrhea. Over-stimulation at the neuromuscular junction causes fasciculations, weakness and paralysis. Over-stimulation within the central nervous system causes altered mental status, seizures, loss of consciousness, and apnea. Death is usually due to respiratory failure. Treatment for organophosphorus nerve agent exposure is with atropine to competitively block the parasympathetic effects and with oximes like pralidoxime to pry off the nerve agent from acetylcholinesterase. The prying of the nerve agent off of acetylcholinesterase must occur before the binding becomes "aged" and thus irreversible.[49]

1. Stewart C. Weapons of Mass Casualties and Terrorism Response handbook. 2006. Jones and Bartlett Publishers, Inc. Boston. Page 2.
2. Hasegawa GR. Proposals for chemical weapons during the American Civil War. Mil Med. 2008 May;173(5):499-506.

3. Russell D, Blain PG, Rice P. Clinical management of casualties exposed to lung damaging agents: a critical review. Emerg Med J. 2006 Jun;23(6):421-4.

4. Meakins JC, Priestley JG. The after effects of chlorine gas poisoning. Can Med Assoc J 1919;9:968-74.

5. Stewart C. Weapons of Mass Casualties and Terrorism Response handbook. 2006. Jones and Bartlett Publishers, Inc. Boston. Page 3.

6. Gas Shell Bombardment of Ypres. London, Public record Office, July 12-13, 1917.

7. Szinicz L. History of chemical and biological warfare agents. Toxicology. 2005 Oct 30;214(3):167-81.

8. Witschi H. Fritz Haber: December 9, 1868-January 29, 1934.

9. Jansen S. chemical-warfare techniques for insect control: insect 'pests' in Germany before and after World War I. Endeavour. 2000;24(1);28-33.

10. Manchester KL. Man of destiny: the life and work of Fritz Haber. Endeavour. 2002 Jun;26(2):64-9.

11. Ellison DH. Handbook of Chemical and Biological Warfare Agents. Second Edition. © 2008 by Taylor & Francis group, LLC. Boca Raton. Page 3.

12. Raffle PAB, Adams PH, Baxter PJ, Lee WR, eds. Hunter's Diseases of Occupations. Eight Edition. 1994. Edward Arnold Publishers. London. Page 261.

13. Hoffman D: Cold war report: Russia's forgotten chemical weapons. WashingtonPost.com; August 16, 1998.

14. Heyndrickx A: chemical warfare injuries. Lancet 1991;337:430.

15. Da R, Blanc PD. Chlorine gas exposure and the lung: A review. Toxicol Ind Health 1993;9:439-55.

16. Beach FXM, Jones S, Scarrow H. Respiratory effects of chlorine gas. Br J Ind Med 1969;26:231-6.

17. Evans RB. Chlorine: state of the art. Lung. 2005 May-Jun;183(3):151-67.

18. Das R, Blanc PD. Chlorine gas exposure and the lung: a review. Toxicol Ind Health. 1993 May-Jun;9(3):439-55.

19. do Pico GA. Toxic gas inhalation. Curr Opin Pulm Med. 1995 Mar;1(2):102-8.

20. Kennedy SM, Enarson DA, Janssen RG, chan-Yeung M. Lung health consequences of report accidental chlorine gas exposures among pulpmill workers. Am Rev respire Dis. 1991 Jan;143(1):74-9.

21. Parrish JS, Bradshaw DA. Toxic inhalational injury: gas, vapor and vesicant exposure. Respir care clin N Am. 2004 Mar;10(1):43-58.

22. Brennan RJ, Waeckerle JF, Sharp TW, Lillibridge SR. Chemical warfare agents: emergency medical and emergency public health issues. Ann Emerg Med. 1999 Aug;34(2):191-204.

23. Garner JP. Some Recollections of Porton in World War I. commentary. JR Army Med corps. 2003 Jun;149(2):38-41.

24. Winder C. The toxicology of chlorine. Environ Res. 2001 Feb;85(2):105-14.

25. Marrs TC, Maynard RL, Sidell FR. Chemical warfare agents: toxicology and treatment. In: Phosgene. John Wiley, 1996:185-202.

26. Diller WF. Medical phosgene problems and their solution. J Occupational Medicine 1978;20:189-93.

27. Stewart C. Weapons of Mass Casualties and Terrorism Response handbook. 2006. Jones and Bartlett Publishers, Inc. Boston. Page 9.

28. Baskin SI: Zyklon, in La Cleur W. (ed): Encyclopedia of the Holocaust. New Haven, CT, Yale University Press, 1998.

29. Casarett & Doull's Toxicology: The Basic Science of Poisons. Fifth Edition. 1996. McGraw-Hill Companies, Inc. New York. Pages 350-1.

30. Kehe K, Szinicz L. Medical aspects of sulphur mustard poisoning. Toxicology. 2005 Oct 30;214(3):198-209.

31. Balali-Mood M, Hefazi M. The pharmacology, toxicology, and medical treatment of sulphur mustard poisoning. Fundam Clin Pharmacol. 2005 Jun;19(3):297-315.

32. Geraci MJ. Mustard gas: imminent danger or eminent threat? Ann Pharmacother. 2008 Feb;42(2):237-46.

33. Dacre JC, Goldman M. Toxicology and pharmacology of the chemical warfare agent sulfur mustard. Pharmacol rev. 1996 Jun;48(2):289-326.

34. Watson AP, Griffin GD. Toxicity of vesicant agents scheduled for destruction by the chemical stockpile Disposal Program. Environ health Perspect. 1992 Nov;98:259-80.

35. Cameron GR, Carleton HM, Short RHD. Pathological changes induced by Lewisite and allied compounds. J. Pathol. Bacteriol. 58:411-422(1946).

36. Windholz M, Budavari s, Biumetti RF, Otterbein ES. (eds) The Merck Index. An Encyclopedia of Chemicals, Drugs, and Biologicals. Merck and Co., Rahway, NJ, 1983.

37. Sollman TH. Lewisite. In: A Manual of Pharmacology and Its Applications to Therapeutics and Toxicology, 8th ed. (T.H. Sollman, ed.), W.B. saunders Co., Philadelphia, PA, 1957, pages 19203.

38. Kehe K, Balszuweit F, Emmler J, kreppel H, Jochum M, Thiermann H. Sulfur mustard research-strategies for the development of improved medical therapy. Eplasty. 2008 Jun 10;8:e32.

39. Sawyer TW, Nelson P. Hypothermia as an Adjunct Therapy o Vesicant-induced Skin Injury. Eplasty. 2008 Apr 30;8:e25.

40. Gates M, Williams JW, zapp JA. Arsenicals. In: chemical Warfare Agents, and Related Chemical Problems—Parts I-II. Summary technical report of Division 9. Washington, DC: National Defense research Committee of the Office of Scientific Research and Development; 1946:83-114.

41. Azizi F, Keshavarz A, Roshanzamir F, Nafarabadi M. Reproductive function in men following exposure to chemical warfare with sulphur mustard. Med War. 1995 Jan-Mar;1191):34-44.

42. Safarinejad MR. Testicular effect of mustard gas. Urology. 2001 Jul;5891):90-4.

43. Fleming CD, Edwards CC, Kirby SD, Maxwell DM, Poter PM, Cerasoli DM, edinbo MR. Crystal structures of human carboxyesterase 1 in covalent complexes with the chemical warfare agents soman and tabun. Biochemisry. 2007 may 1;46(17):5063-71.

44. Wiener SW, Hoffman RS. Nerve agents: a comprehensive review. J Intensive Care Med 2004;19:22-37.

45. Newmark J. The birth of nerve agent warfare: lessons from Syed Abbas Foroutan. Neurology 2004;62:1590-6.

46. Lee EC. Clinical manifestations of sarin nerve gas exposure. JMA 2003;290:659-62.

47. Newmark J. Nerve agents. Neurologist. 2007 Jan;13(1):20-32.

48. Sidell FR, Borak J. chemical warfare agents: II. Nerve agents. Ann Emerg Med. 1992 Jul;21(7):865-71.

49. Cannard K. Te acute treatment of nerve agent exposure. J Neurol Sci. 2006 Nov 1;249(1):86-94.

50. Lopez-Munoz F, Alamo C, Guerra JA, Garcia-Garcia P. The development of neurotoxic agents as chemical weapons during the National socialist period in Germany. Rev Neurol. 2008 Jul 16-31;47(2);99-106.

RADIOLOGICAL AGENTS

"The intense atom glows
A moment, then is quenched in a most cold repose."

Percy Bysshe Shelley

RADIOLOGICAL AGENTS OF terror present a range of threats from thermonuclear devices to dirty bombs. A thermonuclear device is a nuclear bomb. This category includes atomic fission bombs and hydrogen fusion bombs. Atomic fission bombs are the type of nuclear bomb that has actually been used in warfare. The United States dropped atomic bombs on the Japanese cities of Hiroshima and Nagasaki at the end of the Second World War. These bombs operated by causing the release of tremendous energy through the fission or splitting of the uranium isotope 235. Hydrogen bombs are many times more powerful that atomic bombs of the type dropped over Hiroshima. They operate by the initial detonation of an atomic bomb which then generates sufficient heat to cause the fusion or uniting of hydrogen atoms.

The risk of a terrorist group obtaining and delivering a hydrogen bomb appears to be somewhat remote. This is the type of threat that is in the capability of only advanced nation states. However, the possibility of such devices and delivery mechanisms such as ballistic missiles falling into the hands of terrorist organizations is a possibility in the setting of revolutions and other severe social disruptions.

An atomic bomb is a more likely threat. An atomic bomb is easier to produce but is still only likely to be in the capabilities of nation states rather than non-national organizations. However, terrorist detonation of an atomic device could have drastic effects upon society. In fact, atomic device detonations could completely change society. For example, if a suitcase bomb was detonated in a major American city, it would cause a loss of property and lives that make the events of September 11, 2001 appear minor in comparison. There would be a shock that would shake society to the core. However, life tends to go on even after a major calamity. There would be initial panic causing many people to flee major cities for the safety of small cities and rural areas. After the explosion of a nuclear suitcase bomb in New York or London, persons living in a major city such as Chicago or Paris would fear that their city would be the next target of attack. Persons would move their families to stay in second homes or with relatives in smaller cities like Kalamazoo or Peoria. After a few weeks, persons would move back to their large cities and resume their normal routines. The long term impact on civilization would be limited. However, what if there was a second or a third or more nuclear explosion in metropolitan areas. There would be great fear of the next nuclear attack. Large cities would become depopulated and would possibly remain depopulated. Civilization would change forever.

There are three forms of radiation. The first form is x-ray or gamma radiations that are high energy electromagnetic emanations. X-rays are generated by electrons accelerated through an electrical potential striking a metal surface. Gamma radiation is the same as x-rays except they are generated by atomic nuclear decay. X-rays and gamma radiation are electromagnetic waves that are of the same nature as visible light but of higher energy. Gamma and x-rays are very penetrating and shielding requires lead or other high density material.[1]

Beta particles are essentially electrons that have limited penetrating power and can be easily shield by thin aluminum.[1]

Alpha particles are helium atoms which have been striped of their electrons. Alpha particles have high mass but low penetrating power. They may be stopped by skin and can be easily shielded by thin paper.[1]

The main risk of beta and alpha particles is if the materials that emanate these particles are internalized within the body. The most effective means for radioactive beta and alpha emitters to be internalized is through inhalations. If these materials are inhaled, they present a great risk of serious radiation toxicity including cancer in the victim. The main risk of a dirty bomb where radioactive material is spread by detonation of conventional explosives is the inhalation of beta and alpha emitters by persons in the area of the explosion.

1.   Stewart C. Weapons of Mass Casualties and Terrorism Response Handbook. 2006. Jones and Bartlett Publishers, Inc. Boston. Page 184.

LEGAL CONSIDERATIONS

*If the law is upheld only by government officials, then all
law is at an end.*

Herbert Hoover
Message to Congress 1929

HOARDING

THERE MAY BE concerns about hoarding of vital supplies and food stuffs after or before a bioterrorist attack. During past times of crisis there have been efforts by individuals and corporations to store large quantities of supplies. The federal government implemented a program of rationing of foods and consumer goods during the Second World War. Rationing has not occurred to any extent during any military conflict involving the United States since the Second World War. The lack of need for rationing is a manifestation of the large surplus production capacity of the United States compared to the needs of the military during the time of active conflict. However, the collapse of the Soviet Union in the early 1990s occurred at a time of a major economic expansion in the United States. At that time there were thoughts that one of the reasons for the major economic expansion was the "war dividend" due to the lack of further expenditures to fight the "Cold War".

Consequently, there is some evidence that the ability of the United States to maintain production at a level to counter a military treat has some limits even during times of normal economic functioning. The disruptions that would be anticipated to occur during an epidemic of pandemic influenza would be likely to place tremendous stresses upon the supply of foods and consumer goods. The plans of Goodrich Corporation to encourage its employees to stock-up on food and other supplies may run afoul of the plans of the United States federal government. However, it should be noted that the Center for Disease Control website does provide the same recommendations for the general public.

It should be noted that the U.S. Food and Drug Administration (FDA) has authority over foods as well as drugs sold within the United States. The earliest federal legislation that dealt with food and drugs was the Drug Importation Act of 1848 which was passed by the United States Congress in order to require the U.S. Customs Service to conduct inspections so as to stop the importation of adulterated drugs from overseas (Legal Medicine. Fifth Edition. © 2001 by Mosby, Inc. St. Louis, Missouri. Page 560). The first comprehensive Food and Drug Act was enacted by the United States Congress under the Theodore Roosevelt administration on June 30, 1906 (Legal Medicine. Fifth Edition. © 2001 by Mosby, Inc. St. Louis, Missouri. Page 561). In 1938, the United States Congress passed the Federal Food, Drug, and Cosmetic Act (FDCA) (21 United States Code [U.S.C.] 321 to 394) (Legal Medicine. Fifth Edition. © 2001 by Mosby, Inc. St. Louis, Missouri. Page 561).

Each individual state within the United States may have concerns about stresses placed upon local sale and supply of food and consumer goods that may mirror the concerns of the federal government.

During the Second World War persons that were involved in hoarding were often considered to be "Black Marketers".

> *"Black market" was the term given to the illegal trade in consumer goods, manufactured products and raw materials without regard to rationing or price fixing statutes, practiced because of the scarcity of goods. The constantly escalating game of bidding and the risks that the traders ran pushed prices up to unbelievable levels.*

The Historical Encyclopedia of World War II. © 1980 by Facts on File Inc. Greenwich House. New York. ISBN 0-517-431491. Page 57.

In Britain and the United States black marketers were subject to imprisonment. These hoarding activities were a capital offense in totalitarian regimes such as Nazi German and the Soviet Union.

MEDICAL LICENSURE ISSUES

M EDICAL LICENSURE WITHIN the United States is conducted by the various states of the Union. Consequently, a prescription written by a physician in one state may not be deemed valid in another state. The various states in the United States have the right to protect the health of their citizens by means of controlling licensure of physicians to practice medicine within the borders of the State.

> *"The right of a physician to toil in his profession . . . with all its sanctity and safeguards is not absolute. It must yield to the paramount right of government to protect the public health by any rational means."*

Lawrence v. Board of Registration in medicine, 239 Mass. 424, 428, 132 N.E. 174 (1921).

This may result in difficulties in filling of prescriptions in states other than the State where the prescription was written. The licensure statutes of the states had the original propose of ensuring that physicians had the minimum scientific knowledge in order to safety practice (Legal Medicine. Fifth Edition. © 2001 by Mosby, Inc. St. Louis, Missouri. Page 70). However, in the case of *Hawke v. New York, 170 U.S. 189, 194 (1898)* the United States Supreme Court ruled that states have the right to use standards of behavior and ethics as factors in granting or removing physician licensure. Consequently, the various states have the power to control the practice of medicine within their borders through medical licensure. The writing of prescriptions in one state to be filled by the patient in another state may be deemed a clear violation of the right of the State where the prescription is filled to control the practice of medicine in that state.

In addition, during the time of an epidemic there may be migration of persons from colder areas of the United States to warmer areas in the mistaken belief that there is less risk of becoming infected with influenza. As a result, there may be abnormal stresses placed upon the supplies of non-influenza related medications that may be problematic for health delivery. For example, a state such as Florida may experience an abnormally large demand for medication needed to treat diabetes such as insulin. This may result in a lack of necessary medications for diabetics who usually live in Florida. Conversely, there may be an oversupply of insulin in a cold northern state such as Michigan. The federal government may wish to control the supply of vital medical supplies across state borders. This control of the flow of medical supplies across state borders may be controlled by the federal government through the operation of the Interstate Commerce Act. Interstate and foreign commerce are defined as follows:

> *Commerce between a point in one State and a point in another State, between points in the same State, through another State or through a foreign country, between points in a foreign country or countries through the United States and a point in a foreign country or in a Territory or possession of the United States, but only insofar as such commerce takes place in the United States. The term "United States" means all the States and the District of Columbia.*

18 U.S.C.A. § 831.

Black's Law Dictionary. Fifth Edition. © 1979 West Publishing Co. St. Paul Minn. ISBN 0-8299-2041-2. Page 735.

Interstate commerce was defined in *Gibbons v. Ogden, 22 U.S. (9 Wheat.) 1, 6 L.Ed. 23* as follows:

> *Traffic, intercourse, commercial trading, or the transportation of persons or property between or among the several states of the Union, or from or between points in one state and points in another state; commerce between two states, or between places lying in different states.*

Black's Law Dictionary. Fifth Edition. © 1979 West Publishing Co. St. Paul Minn. ISBN 0-8299-2041-2. Page 735.

Furst v. Brewster, 282 U.S. 493, 51 S.Ct. 295, 296, 75 L.Ed. 478 further defines interstate commerce as "It comprehends all the component parts of commercial intercourse between different states." (Black's Law Dictionary. Fifth Edition. © 1979 West Publishing Co. St. Paul Minn. ISBN 0-8299-2041-2. Page 735.)

The Interstate Commerce Act provides for the following:

> *The act of congress of February 4, 1887 (49 U.S.C.A. § 1 et seq.), designed to regulate commerce between the states, and particularly the transportation of persons and property, by carriers, between interstate points, prescribing that charges for such transportation shall be reasonable and just, prohibiting unjust discrimination, rebates, draw-backs, preferences, poling of freights, etc., requiring schedules of rates to be published, establishing a commission to carry out the measures enacted, and prescribing the powers and duties of such commission and the procedure before it.*

Black's Law Dictionary. Fifth Edition. © 1979 West publishing Co. St. Paul Minn. ISBN 0-8299-2041-2. Page 735.

Clearly the federal government has the power to have some degree of control over the flow of medical supplies and pharmaceuticals across state borders.

However, the federal government may not be able to control the flow of persons across state borders due to the privileges and immunity clause of the United States Constitution. Perhaps more importantly, the various states may not normally control the flow of United States citizens into their state from other states. The Privileges and Immunities Clause of the United States Constitution provides the following protections:

> *There are two Privileges and Immunities Clauses in the Federal Constitution and Amendments, the first being found in Art. IV, and the second in the 14th Amendment, § 1, second clause, clause 1. The provision in Art. IV states that "The Citizens of each State shall be entitled to all Privileges and immunities of Citizens in the several States," while the 14th Amendment provides that "No state shall make or enforce any law which shall abridge the privileges or immunities of citizens of the United States.*

> *The purpose of these Clauses is to place the citizens of each State upon the same footing with citizens of other states, so far as the advantages resulting from citizenship in those states is concerned.*

Black's Law Dictionary. Fifth Edition. © 1979 West publishing Co. St. Paul Minn. ISBN 0-8299-2041-2. Page 1079.

Consequently, there is likely to be a bar against actions by the governments of the various states in attempting to prevent entry of United States citizens into their state from another state. This bar may be effective even during the time of a crisis such as an avian influenza epidemic.

The various states within the United States regulate the prescriptive authority of the physicians, dentists, and nurse practitioners within each state. Consequently, one state may have a different law concerning prescriptive authority than another state. For example, one state may allow a nurse practitioner to write prescriptions while another state may not allow nurse practitioners to write prescriptions. As a result, a prescription written by a nurse practitioner in a state granting prescriptive authority to nurse practitioners will almost certainly not be honored in a state that does not grant nurse practitioners prescriptive authority. In addition, even if a practitioner such as a physician is of a profession that is granted prescriptive authority in two different states, the writing of a prescription by a physician licensed in one state with knowledge that the patient intends to fill the prescription in another state in which the physician is not licensed may be considered the unauthorized practice of medicine. It should be noted that the unauthorized practice of medicine is usually considered a felony.

The unset of an epidemic may lead people to attempt to obtain medical advice and treatment in means other than through direct contact. There are likely to be attempts to obtain medical advice and treatment via electronic devices such as the telephone or the internet. The reasons for the desire to obtain medical treatment over a distance through electronic means are likely to be three fold. First, there will be reluctance by persons to attend physician offices where there will be an increased likelihood of coming into contact with sick persons who may be infected with avian influenza. There may also be a reduction in the availability of physician services within the patient's immediate geographic area. This will cause patients to seek medical advice and treatment at a distance. The providers of this medical advice and treatment may be located across state borders. In addition, patients may seek the advice of specialist with particular expertise in the diagnosis and treatment of disease caused by avian influenza. These considerations will encourage patients to seek medical care from physicians that may not be licensed to practice medicine within the state of residence of the patient. Physicians may also be drawn to provide care to patients at a distance who may not be located within the same state as the physician and who may have never been seen by the physician. The factors that will draw physicians into this telemedicine activity are also three fold. First, there may be commercial factors that may draw physicians into areas of practice in great demand. Due to the demand, these areas of specialized practice may be very lucrative. Physicians may also wish to provide medical care without the risk to their personal health due to infection with avian influenza that would arise with direct patient contract. In addition, there will be the admirable desire on the part of physicians to provide needed services to persons that may not have access to medical services in their own communities due to remote geography or other factors limiting access to care. The issue of telemedicine arise a number of legal concerns.

The first legal consideration concerning telemedicine services involves legal liability for medical malpractice. The case of *International Shoe v. State of Washington, 325 U.S. 310, 66 S.Ct. 154 (1945)* established the principle that in order for a state to have legal jurisdiction over a person, that person must have some minimal contacts with the state seeking the jurisdiction. Within the realm of medical malpractice, it is generally considered that the physician can only be sued within the state where he provided the medical service even if the patient has a place of residence in another state (Legal Medicine. Fifth Edition. © 2001 by Mosby, Inc. St. Louis, Missouri. Page 238). However, in *Bullion v. Gillespie, 895 F. 2d 213 (5th Cir. 1990)* a physician was held to be subject to the jurisdiction of another state where the patient lived when it was shown that the physician attracted patients to his practice through national distributed marketing literature. In *Kenndy v. freeman, 919 F. 2d 126 (10th Cir. 1990)* a physician was found to be subject to the jurisdiction of another state when he was involved in the reading of biopsy specimen sent by physicians located outside of his own state. It should also be noted that several states have forbidden the prescription of medications for patients where the only contact with the patient was through telemedicine (Legal Medicine. Fifth Edition. © 2001 by Mosby, Inc. St. Louis, Missouri. Page 238). *In re B.T. Taylor, M.D., Action report, Medical board of California (Oct 1999)*, the Board of Medicine of the State of California reprimanded a California physician when it received notification of a disciplinary action by the State of Colorado due to the physician having prescribed medications to patients in Colorado with whom the physician only had contact via the internet.

Each national government has the power and right to regulate the practice of medicine within its national borders. Prescriptions written in foreign countries are not likely to be honored in a nation other than the country of origin of the prescription. The public policy issues in this matter are similar to the concerns of the various states of the United States in preventing the unauthorized practice of medicine. It should be noted that the United States Supreme Court only ruled in 1973 that a requirement that a person be a citizen of the United States in order to obtain medical licensure is an unconstitutional discrimination (Legal Medicine. Fifth Edition. © 2001 by Mosby, Inc. St. Louis, Missouri. Page 71). It should not be surprising if foreign nations may consider the medical advice to its citizens or residents by a physician that is not a citizen of that nation is an unauthorized practice of medicine.

MEDICAL DECISION MAKING

I N WESTERN DEMOCRACIES including the United States there is the understanding that persons have the right to make their own decisions about matters influencing their health. This concept is termed *autonomy* from the Greek *auto nomos* which means self-rule. This means that in a medical setting that a patient has freedom of choice. (Legal Medicine. Fifth Edition. © 2001 by Mosby, Inc. St. Louis, Missouri. Page 291). In *Schoendorf v. Society of New York Hospital, 1914, 105 N.E. 92 (N.Y.C.A.)* Justice Cordozo stated that "Every human being of adult years and sound mind has a right to determine what shall be done with his own body."

In 1948 there was a Universal Declaration of Human Rights that were ultimately adopted by the World Medical Association and the World Health Organization. Their measures were designed to prevent some of the great evils that occurred during World War II. The focus was to prevent human experimentation of the type that occurred during the Nazi regimen (Legal Medicine. Fifth Edition. © 2001 by Mosby, Inc. St. Louis, Missouri. Page 291).

POLICE POWERS

THE HEALTHCARE SYSTEM and public health systems differ in that public health systems have "police powers" while healthcare systems do not have "police powers".

Police powers are those powers that are necessary to function in a police capacity. The police powers of local public health officials in the State of Michigan in the event of an epidemic, emergency or case where quarantine is required are described in the Public Health Code of the State of Michigan (Act 368 of 1978, as amended) as follows:

> *333.2453 Epidemic; emergency order and procedures; involuntary detention and treatment.*
>
> *Sec. 2453. (1) If a local health officer determines that control of an epidemic is necessary to protect the public health, the local health officer may issue an emergency order to prohibit the gathering of people for any purpose and may establish procedures to be followed by persons, including a local governmental entity, during the epidemic to insure continuation of essential public health services and enforcement of health laws. Emergency procedures shall not be limited to this code.*
>
> *(2) A local health department or the department may provide for the involuntary detention and treatment of individuals with hazardous communicable disease in the manner prescribed in sections 5201 to 5238.*

The difference between a public health department and a hospital as part of a healthcare system is illustrated by what happens in the case of a person with a serious infectious disease such as tuberculosis. A hospital that has a patient with infectious tuberculosis who refuses to take anti-tubercular medication cannot force the person to take the medication. If the person with infectious tuberculosis wishes to leave the hospital and enter into the general community with the risk of spreading tuberculosis, the hospital must allow the infectious person to leave since the hospital is a non-public organization without police powers. In this situation the hospital must contact the local public health department which then exercises its police powers to detain the infectious individual and compels the person to take anti-tubercular medications.

The essence of the above dissertation is that the healthcare and public health systems are fundamentally different. The focus of the healthcare systems is on provision of healthcare services to individuals. The focus of the public health systems is detection and control of disease in populations. Public health systems have police powers. Healthcare systems do not have police powers. There is a fundamental difference in the training and therefore the mind set of physicians working in public health compared to physicians working in healthcare.

Police powers are those powers that are necessary to function in a police capacity. The police powers of local public health officials in the State of Michigan in the event of an epidemic, emergency or case where quarantine is required are described in the Public Health Code of the State of Michigan (Act 368 of 1978, as amended) as follows:

> *333.2453 Epidemic; emergency order and procedures; involuntary detention and treatment.*
>
> *Sec. 2453. (1) If a local health officer determines that control of an epidemic is necessary to protect the public health, the local health officer may issue an emergency order to prohibit the gathering of people for any purpose and may establish procedures to be followed by persons, including a local governmental entity, during the epidemic to insure continuation of essential public health services and enforcement of health laws. Emergency procedures shall not be limited to this code.*
>
> *(2) A local health department or the department may provide for the involuntary detention and treatment of individuals with hazardous communicable disease in the manner prescribed in sections 5201 to 5238.*